100 Questions (and Answers) About Statistics

100 Questions (and Answers) About Statistics

Neil J. Salkind
University of Kansas

Los Angeles | London | New Delhi
Singapore | Washington DC

Los Angeles | London | New Delhi
Singapore | Washington DC

FOR INFORMATION:

SAGE Publications, Inc.
2455 Teller Road
Thousand Oaks, California 91320
E-mail: order@sagepub.com

SAGE Publications Ltd.
1 Oliver's Yard
55 City Road
London, EC1Y 1SP
United Kingdom

SAGE Publications India Pvt. Ltd.
B 1/I 1 Mohan Cooperative Industrial Area
Mathura Road, New Delhi 110 044
India

SAGE Publications Asia-Pacific Pte. Ltd.
3 Church Street
#10–04 Samsung Hub
Singapore 049483

Printed in the United States of America

Library of Congress Cataloging-in-Publication Data

Salkind, Neil J., author.

100 questions (and answers) about statistics / Neil J. Salkind, University of Kansas.

pages cm

Includes bibliographical references and index.

ISBN 978-1-4522-8338-8 (pbk.)
ISBN 978-1-4833-1314-6 (web pdf) 1. Statistics—Miscellanea.
I. Title. II. Title: 100 questions about statistics. III. Title: One hundred questions (and answers) about statistics.

QA276.12.S235 2015

519.5—dc23   2014005753

This book is printed on acid-free paper.

Acquisitions Editor: Vicki Knight
Editorial Assistant: Yvonne McDuffee
Production Editor: Libby Larson
Copy Editor: Paula L. Fleming
Typesetter: C&M Digitals (P) Ltd.
Proofreader: Sally Jaskold
Indexer: Terri Corry
Cover Designer: Candice Harman
Marketing Manager: Nicole Elliott

Certified Chain of Custody
SUSTAINABLE FORESTRY INITIATIVE
Promoting Sustainable Forestry
www.sfiprogram.org
SFI-01268

SFI label applies to text stock

14 15 16 17 18 10 9 8 7 6 5 4 3 2 1

Contents

Preface

It seems that almost everywhere we turn today, the term *big data* is being used for explaining everything from the economy's success or failure to predicting who will win the World Series. Understanding quantitative information and its patterns is more important than ever, and aspiring professionals, students, policy makers, and those in thousands of other professions need the how-to of understanding data—hence, the importance of basic and advanced statistics.

100 Questions (and Answers) About Statistics is an attempt to summarize the most important questions about basic statistics, from a simple understanding of summary statistics to more sophisticated inferential statistics.

100 Questions (and Answers) About Statistics was written because I have noticed in my own years of teaching that many students and professionals need a short review to steer them in the right direction as to what topics to focus on and where to look for further information.

This short book is intended for those individuals who need a refresher on what the important topics are within statistics, as well as for those who are entirely new to the discipline and need a resource to discover the key questions one might ask. Think of *100 Questions . . .* as a reminder, a resource, and a refresher. It's for graduate students preparing for comprehensive exams, researchers who need a reference, undergraduates in affiliated programs who will not be taking a primary course in statistics, policy makers who need to know about these techniques, and anyone who is simply curious about how these tools can most effectively be used.

Some tips about using the book:

1. The questions are divided into 12 parts as follows:

   - Part 1: Why Statistics?
   - Part 2: Understanding Measures of Central Tendency
   - Part 3: Understanding Measures of Variability
   - Part 4: Illustrating Data
   - Part 5: Understanding Relationships
   - Part 6: Understanding Measurement and Its Importance
   - Part 7: Understanding the Role of Hypotheses in Statistics
   - Part 8: Understanding the Normal Curve and Probability
   - Part 9: Understanding the Concept of Significance
   - Part 10: Understanding Differences Between Groups

- Part 11: Looking at Relationships Between Variables
- Part 12: Other Statistical Procedures

2. Each of these questions and answers can stand alone as a direct question and relatively short answer. Probably a book-length response could be written for each question, but the task at hand here is quickly accomplished—provide enough information to allow the reader to gain some knowledge before he or she moves on to the next question or topic.

3. Not all these questions and answers are independent of one another; in fact, most supplement each other. This is to help reinforce important material and make sure that both primary and secondary topics get consideration.

4. Each question ends with reference to three other, related questions— the three that we think can best supplement the primary question. There are of course, many other questions related that could be listed.

Acknowledgments

Thanks go to the best editorial and production teams in the business—especially to Vicki Knight, who enthusiastically greeted the idea for this series. She tolerated an unbelievable number of late-night emails from me with grace and provided support and great assistance. To assistant editors Lauren Habib and Jessica Miller: They made things happen quickly and efficiently and have been a huge help. Thanks also to production editor Libby Larson and copy editor Paula L. Fleming. I own all the errors here. Please let me know via email how this book might be better.

Thank you to the following reviewers for their time and input: T. John Alexander, Texas Wesleyan University; Jamie V. Brown, Mercer University; Yeonsoo Kim, University of Nevada, Las Vegas; Linda Martinez, California State University–Long Beach; Juliana Raskauskas, California State University–Sacramento; Jennifer R. Salmon, Eckerd College; and Robin L. Spaid, Morgan State University.

And to the best kids in the galaxy: Sara, Micah, and Ted and—of course—Lucky, Pepper's heir apparent, and Lew M., still after all these years brave, indefatigable, caring, courageous, and honest.

Neil J. Salkind
Lawrence, Kansas
njs@ku.edu

About the Author

Neil J. Salkind is Professor Emeritus at the University of Kansas where he taught in the Department of Educational Psychology for more than 35 years. His interest is in writing about statistics and research methods in an informative, nonintimidating, and noncondescending way. He is also the author of *Statistics for People Who (Think They) Hate Statistics, Fifth Edition*; *Statistics for People Who (Think They) Hate Statistics (Excel 2010 Edition), Third Edition*; *Excel Statistics—A Quick Guide, Second Edition*; and *100 Questions (and Answers) About Research Methods* (first in the SAGE 100 Questions and Answers series), and editor of the *Encyclopedia of Research Design*.

PART 1

WHY STATISTICS?

What Is the Study of Statistics, and Why Is It Important?

We all live in a world that is increasingly dominated by data. Who would have thought that the analysis of data would be as important to professional sports teams as it is to educational institutions and Fortune 500 businesses? Yet, it is. With increasing frequency, people are turning to the discipline of statistics to look for patterns and predict outcomes. The study of statistics contains a set of tools that helps us better understand complex outcomes and make decisions.

Statistics is the description, organization, analysis, and interpretation of quantitative information. This information might be a set of test scores, a preference for a particular type of automobile, or how often a basketball team scores after a steal from the other team. You can answer questions such as these, plus almost any other question that deals with the analysis of data, by using the various tools that you will learn about throughout this book.

The study of statistics is important for several reasons. In the most applied way, it helps us make decisions based on information otherwise too difficult or impossible to interpret. This usefulness can be apparent in even the most simple of cases. For example, rather than individual test scores for a group of students, wouldn't the average score for all the students be more helpful? Or, wouldn't you rather work with the average results of a survey about how a group of customers felt about the service they were given rather than each customer's response to 20 different questions?

These two examples point to the fact that being able to collect, describe, and analyze information leads us to make better decisions because our decisions are based on evidence. And the tool that allows us to explore what that previously unorganized set of information might mean? Statistics!

Statistics is invaluable to the research that scholars do, the decisions that local and national politicians make, and even the everyday functioning of businesses that have to act on information to further their goals.

More questions? See questions #2, #3, and #4.

How Did Statistics "Get Started"?

The study of statistics goes way beyond the collection and analysis of data—it's much more about the collection and use of *information* to make important decisions. There has probably never been a time when people have not been concerned with how many of something they had (such as "How much food do we have until we run out?" or "How many days until winter?") and how those numbers affected certain outcomes (such as well-being and shelter).

So from the start, numbers were attached to particular outcomes. If one did well in school and got good grades, there was a higher likelihood of success in future classes. If one got a good education, then a better job might await upon graduation. And it was not that long ago that those whom we know as demographers today (people who study populations and their characteristics) started counting and looking at distributions of where the most people lived, worked, and played.

All of this was mostly done by mathematicians, but as disciplines such as biology and, later, psychology were pressed for an understanding of what was being observed, the field of statistics was born.

Probably a major milestone in that birth was the work of Francis Galton, a first cousin of Charles Darwin who was born in the early 19th century. Galton invented the still very popular tool called the correlation coefficient, which looks at the relationship between variables. His interest? Intelligence among families. His work (though often questioned later on) laid the framework for comparing such relationships among family members.

After Galton, statistics saw a ton of new developments as an increasingly complex society increased demand to understand the complexity of all the information that was available. Such names as Karl Pearson (mathematician) and R. A. Fisher (agronomist) applied what they learned from their own fields of study and to different aspects of human behavior. With the advent of personal computers over the last 40 years, the most powerful of statistical techniques have become available to almost anyone who might want to look at patterns and trends in large data sets—a very important part of modern-day statistics. Even college and professional sports teams now use this approach to identify what works—and what doesn't.

You can find more about the history of statistics at Saint Anselm College's website: www.anselm.edu/homepage/jpitocch/biostatshist.html.

More questions? See questions #1, #3, and #4.

What Are Descriptive Statistics, and How Are They Used?

Descriptive statistics are used to organize and describe the character-istics of a set of data (also called a distribution). Descriptive statistics are the first tools used to explore the data, getting some important indications of what the data set "looks like."

There are two general categories of descriptive statistics.

The first category consists of those that look at measures of central tendency, such as the mean, the median, and the mode—all of which are referred to as "averages." Any one of these can be used to represent the one best point that represents a set of data. For example, if we were interested in hamburger sales at a local restaurant, the first question we might ask is how many burgers are sold, on average, each week. As another example, we might ask, "What's the most frequently sold model of Ford cars?"

The second category consists of those descriptive tools that look at variability or spread or dispersion among a set of data. The range, standard deviation, and variance are among these measures. These measures tell us how much the data points in a set differ from one another. For example, if we want to know how similar a group of children are in their reading skills, we might look at the standard deviation for their reading-test scores. The lower the value of that descriptive statistic, the less variable (and the less different) these children are from one another.

Combined, these two categories—averages and measures of variability—give us a very good picture of the characteristics of a data set and how it might differ from other data sets. And, as you will see, these two types of measures are the basis of more sophisticated statistical operations, such as looking at the significance of the difference between the averages of two groups of observations.

More questions? See questions #4, #7, and #16.

What Are Inferential Statistics, and How Are They Used?

Inferential statistics allow us to infer findings from a smaller group (often called a sample) to a larger group (often called a population). Inferential statistics are often an extension of descriptive statistics, but this is not always the case. Sometimes, knowing the average of a set of scores (using descriptive statistics) is sufficient, given the question being asked, and it is not necessary to use inferential statistics to see, for example, whether a particular average of one group differs from the average of another group.

As an example of how inferential statistics could be used, consider the examination of differences between two groups of elementary school students on a test of reading skills. One group might get extra instruction, while another group would not. The two samples would then be compared, and the results would be generalized to the population from which these two samples were selected.

Why not just test the entire population of elementary school children? In most cases, populations are too large—the expense in both time and dollars is too great. Scientists have learned enough about statistics over the past 200 years that they can very accurately assess how well a much smaller group of observations, based on a sample, represents the larger population and use that information to make judgments about the population.

As you might imagine, how well a sample represents the population is a key element in the use of inferential statistics. Representativeness helps to determine whether the results of a study can be generalized from the sample to the population. A great deal of the accuracy of the results of an inferential procedure depends on how well the sample is selected from the population; the more representative the sample is of the population, then the more confidence one can have in inferring from the sample to the population, and the higher is the likelihood that the results are generalizable to the population (and to other similar populations as well).

Inferential statistics such as *t*-tests between means, analysis of variance, and regression form part of the basis of most introductory statistics courses and are very widely used. In sum, descriptive statistics allow us to describe the characteristics of a set of data, while inferential statistics allow us to apply those observations to larger groups.

More questions? See questions #60, #70, and #77.

I'm Not Going to Be a Statistician—Why Should I Be Taking a Course in Statistics?

This answer goes beyond the most obvious one—that you may need to take a statistics course to fulfill degree requirements, or you are studying for a major exam and part of your knowledge is supposed to be in this area.

Actually, there are at least four other good reasons for taking a course in statistics.

First, this is a challenging and intellectually interesting topic that you may not have experienced in your previous educational efforts. It is probably a discipline that you have not yet encountered, and it can challenge you and help stretch your skills—always a praiseworthy endeavor.

Second, in a society that increasingly bases decisions on data and other evidence, an understanding of statistics is invaluable. It will prepare you to adeptly understand patterns and trends in the data and use them to make informed decisions. You will be the one who asks whether data are available to understand which medical treatment might be best, why one way of approaching math instructions is better than another, or whether electronic textbooks are more effective than print ones. Statistics is a set of tools that shows you how to pose and answer any research-based question you might want to ask.

Third, you will be a better-informed citizen who is ready to interact intelligently and capably with your fellow students, your teachers, and others. When the results from a journal article are discussed or when a conclusion is offered regarding an interesting social, medical, or behavioral (among others) outcome, you will understand what is being said and be able to respond perceptively and in an informed way.

Finally, statistics coursework is a terrific way to prepare for further educational and professional opportunities such as graduate work or, if you are already a graduate student, future work activities. All done with your schooling? Then understanding statistics gives you more potential for advancement in your professional field.

More questions? See questions #1, #2, and #6.

What Statistical Software Packages Should I Consider Using? SPSS? Excel? Others?

Many statistical analysis software packages are available; some are free, some are shareware (you pay what you think the software is worth), and some are commercial products. Which one you choose will depend upon several factors.

SPSS, a very popular software package has been produced for many years, is used mostly by researchers in the social and behavioral sciences. It is expensive, but most postsecondary institutions have installed it on their campus-wide servers. Therefore, if you are a student, you can use it without additional cost. Plus, relatively inexpensive student versions are available, though these are limited in capability.

Excel, the world's most popular spreadsheet program, allows you to manipulate rows and columns of numbers and text. While it is not a dedicated statistical analysis program, it contains so many functions and built-in analytic tools that it can almost replace products such as SPSS. Excel also has powerful graphing tools, and it is part of the Microsoft Office package, which contains Word, so inserting the results from an Excel analysis in a Word document is very easy. As with SPSS, many institutions have Excel readily available. In *100 Questions (and Answers)*, we will use Excel for our examples because it is the most readily available and easiest-to-use software for most beginning and intermediate users.

What should you choose for your work? Either SPSS or Excel will work for almost anything discussed in this book (as will many other less popular programs). However, keep the following questions in mind as you decide whether you should use one of these or another package:

- How expensive is the program? Is it available at no charge to you through your institution?
- Is there a student version, and are you eligible to purchase it?
- Is the program modular, and if so, what modules might you need?
- Is technical support by either phone or email readily available?

- Do you have the correct hardware (memory, storage space, etc.) to use the software?
- Does your computer have a compatible operating system to run the software?

Be sure the software fits your needs before you commit to buying it, but remember: Almost any program will meet basic needs.

More questions? See questions #16, #79, and #80.

PART 2

UNDERSTANDING MEASURES OF CENTRAL TENDENCY

What Are Measures of Central Tendency, and Why Are They Used?

Measures of central tendency such as the mode, the median, and the mean are the first type of descriptive statistics that one uses to describe the characteristics of a set of data. The second type of statistic, measures of variability, will be discussed in the next section in *100 Questions*.

Measures of central tendency, also called averages, reflect the one most central or representative point in a set of data. In other words, if you had to select only one score from a set of scores to represent that set, you would chose a measure of central tendency. For example, if you needed to best represent the car sales at the local Honda dealership for the month of August, you might use the mean as the descriptive statistic. The specific measure of central tendency one uses depends on the type of data one is dealing with.

The median is the score that is the midpoint among a group of scores when the scores are put in order from least to greatest. It is defined as the point at which 50% of the scores are above it and 50% of the scores are below it. It is most often used to compute the average of a set of scores when the set includes extreme (very high or very low) scores.

The mode, which is the most imprecise measure of central tendency, is the number of times that the most frequently occurring score appears in a set of data. It is most often used when the data are nominal or categorical in nature.

The mean is the arithmetic average of a set of scores. While there are different types of means, the arithmetic mean is the point at which the set of data is "balanced" or the point at which the various scores find a middle ground.

While each of these measures of central tendency is used for different types of data, one of the most important applications is to describe a data set along with measures of variability and as the basis for the computation of standard scores. In addition, an interesting applied question when dealing with any type of data is, What is the relationship between an "average" score and scores that are "not average" or atypical?

More questions? See questions #9, #11, and #13.

What Is an Example of How a Measure of Central Tendency Can Be Used?

Averages are often used as a reference point for one characteristic of a set of scores, such as the average heights of males and females. Averages are the "middle" ground: The average is the best representation of all the scores in a set of scores. In both inferential as well as descriptive statistics, averages are widely used as measures that can be compared to one another to determine whether there is any difference between two or more groups.

One such study examined the perceptions of love across the life span using a triangular theory of love, which distinguishes among passion, intimacy, and commitment. The study investigated the results of tests in which the short Triangular Love Scale was administered to adolescents and adults, and it tracked age and gender differences in a sample of almost 3,000 12- to 88-year-olds.

Where do averages fit in? In answering their questions, the researchers used an inferential test—called the *t*-test—that compares the mean score of one group of participants to the mean score of another group of participants. *t*-tests, which use the arithmetic mean as the average (along with a measure of variability), enable the researchers to reach a conclusion as to whether the difference between groups is due to chance or due to the variable of interest. In this case, the variable of interest was the score on the Triangular Love Scale.

The results? Adolescents (ages 12–17 years) reported lower levels of all love components compared to young adults (ages 18–30 years), and older adults (50+ years) reported lower levels of passion and intimacy but similar levels of commitment compared to young and middle-aged adults (ages 30–50 years). Gender differences in the perceptions of all three love components were present but less sizeable than the researchers expected.

The use of the mean as an indicator of differences and the *t*-test (which comes in many forms) is very common in basic statistics.

Here's the complete reference . . .

Sumter, S. R., Valkenburg, P. M., & Peter, J. (2013). Perceptions of love across the lifespan: Differences in passion, intimacy, and commitment. *International Journal of Behavioral Development, 37*(5), 417–427.

More questions? See questions #74, #82, and #83.

What Is the Mean, and How Is It Computed?

The mean is the most commonly used measure of central tendency. It is the sum of all the values in a group divided by the number of values in the group. A more technical definition is that the mean is the point about which the sum of the deviations equals zero. The type of mean we are discussing here is also called the arithmetic mean.

The formula is quite simple, as follows.

$$\overline{X} = \frac{\Sigma X}{n}$$

where

$\overline{X}$ equals the mean,

ΣX is the sum of all the scores, and

n is the sample size.

Note: In statistics, a lowercase n is used to represent the sample size, and an uppercase N is used to represent the population size.

To compute the mean, follow these steps:

1. List all the values in the data set.

2. Sum the values.

3. Divide by the number of observations.

For example, the set of scores 7, 8, 4, 6, and 5 totals 30, and the mean is 30/5 or 6.

The mean is most often represented by an uppercase X with a bar over it, but it is often represented using M.

One caution about using the mean as a measure of central tendency is that it is not sensitive to extreme scores. For example, the mean of the scores 4, 6, 7, 8, and 20 is 9, but this number does not best represent the set of scores since the average is pulled toward the extreme data point of 20.

More questions? See questions #7, #15, and #16.

What Is an Example of How the Mean Can Be Used?

Over the next 50 years, the United States is going to be a "minority-majority" society, in which the most common racial group will constitute a plurality but not a majority of the population. Therefore, it is important for the social and behavioral sciences to learn more about once-minority populations such as Latinos, Asians, and African Americans.

The researchers Christopher Ellison, Nicholas Wolfinger, and Aida Ramos-Wada studied the importance of the rapid growth of the Latino population in the United States and the role that Catholicism plays in their attitudes toward the family, even though nearly one third of Latinos are not Catholic. A common practice among researchers is to use data that are already available. In this case, data from the 2006 National Survey of Religion and Family Life were used. This survey of working-age adults (ages 18–59 years) in the lower 48 states explored the relationship among multiple dimensions of religiosity—denomination, church attendance, prayer, and beliefs about the Bible—and Latinos' attitudes regarding marriage, divorce, cohabitation, and casual sex.

The findings were that, compared with Catholics, evangelical Protestants tend to hold more conservative attitudes on family-related issues. Latinos who attend services regularly and pray frequently also report more traditional views. Analysis of the data suggests that religious variables are just as potent as socioeconomic and demographic factors in explaining individual-level variation in Latinos' attitudes.

To reach these conclusions, the analysis included an examination of means. The partial results are shown in Table 10.1. Here you can see the attitude items (e.g., "Casual sex is OK") and the mean scores for all respondents, Catholics, evangelicals, others, and those with no affiliation. These means were then subjected to additional analysis, but this is a good example of how averages can be used to assess attitudes.

Table 10.1 How some descriptive statistics are reported

	All	Catholic	Evangelical	Other	No Affiliation
Dependent variables					
Casual sex is OK	3.25 (765)	3.18 (483)	3.59 (147)	3.46 (35)	2.98 (80)
T test (two-tailed)	—	—	2.55*	0.91	−0.99
Loveless couples should not divorce	2.24 (784)	2.03 (489)	2.90 (154)	2.81 (36)	1.80 (84)
T test (two-tailed)	—	—	6.43***	3.14**	−1.41

Here's the complete reference . . .

Ellison, C. G., Wolfinger, N. H., & Ramos-Wada, A. I. (2013). Attitudes toward marriage, divorce, cohabitation, and casual sex among working-age Latinos: Does religion matter? *Journal of Family Issues, 34*(3), 295–322.

More questions? See questions #7, #9, and #16.

What Is the Median, and How Is It Computed?

The median is another measure of central tendency and another type of average. It is the point in a distribution of data where 50% of the cases fall above it and 50% of the cases fall below it. The median is most often represented using the abbreviation *Md*, but it sometimes appears as *M*.

To compute the median, follow these steps:

1. List all of the numbers in descending order, with the largest value at the beginning or top. Here's an example:

   87, 72, 65, 45, 23

2. Select the score that is the midpoint of the set of scores. In this case, it is the third score (with two scores below it and two above it), and the median is 65.

If the set has an even number of scores (such as 87, 72, 67, 65, 45, and 23), the median is the average of the two midpoint scores. In this case, there are six scores, and the two midpoint scores (the third and fourth score) are 67 and 65. The median is the average of those two, or 66.

The median is an interesting and useful measure of central tendency for one reason: It is insensitive to extreme scores, giving a much clearer picture than the mean of the true average of a set of scores that includes an extreme value. Remember, we want an average—whether the mean, median, or mode—to be the one most representative score from a set of scores.

For example, look at the following set of scores (which has one extreme score):

107, 30, 28, 25, 24

The mean is 40 (200/5), and the median is 28. The tug of the one extreme score (107) pulls the average way up. The median of 28, on the other hand, better approximates the most representative score from this set. Using the median moderates the extreme score's effect.

Finally, the median is all about the score that corresponds to the number of cases in the set of scores and not the value of those scores. It is always the 50th percentile or the point that bisects the distribution of scores.

More questions? See questions #7, #12, and #16.

What Is an Example of How the Median Can Be Used?

Y ou already know from the previous question that the most useful application of the median as a measure of central tendency is to data sets that include extreme scores. As it turns out, one of the most frequent uses of the median is the computation of a measure of central tendency for annual income in the United States and in other developed countries.

The reason for this is that the annual income in these countries is widely varied, as reflected by the range of values and extreme scores.

For example, take the following seven annual salaries listed in descending order from lowest to highest.

$235,495

$60,100

$58,768

$54,675

$47,698

$45,687

$41,675

You remember that a measure of central tendency is the one score that best represents the set of all scores. Upon examination of these scores, you can see that all but one are within the range of $41,000 to $61,000. So, to reflect the most central point, you would expect a measure of central tendency to be around $50,000.

The mean of these scores is $77,728, which points to a much higher average then a cursory examination of the data warrants. The median, on the other hand, is $54,675—just what we would expect based on where the majority of values fall. It's that one high value of $235,495 that throws off the mean as a useful measure of central tendency and calls for the median to be used.

How would one know to use the median rather than the mean? Two ways. First, through a visual examination of the data. And, second, when data are ordinal in nature (such as rank in class, largest to smallest), the median also seems to be a more accurate reflection of the central-most point than is the mean.

More questions? See questions #7, #11, and #16.

What Is the Mode, and How Is It Computed?

The mode is the value that occurs most frequently in a group of values, and it is computed by counting the number of times a value appears and identifying the one that occurs most frequently.

For example, look at the following set of values. Here we are defining a value as the label attached to a particular outcome—in this case, different colors.

Red	Blue	Blue	Gray	Violet
Yellow	Blue	Blue	Gray	Violet
Yellow	Blue	Blue	Gray	Violet
Yellow	Blue	Gray	Violet	Violet
Blue	Blue	Gray	Violet	Violet

If you count the frequency that each value appears, you end up with the following:

Value	Frequency
Red	1
Yellow	3
Blue	9
Gray	5
Violet	7

The most frequently occurring value is Blue (which appears nine times), so the mode is the value Blue.

One of the most common errors made regarding the mode is reporting the mode as the frequency with which a value occurs rather than the value itself. So, in the above example, the mode is not 9 but rather is the value that appears most often, or Blue.

One important thing to remember about the mode is that if all the values in a set of data occur with equal frequency, then that set of data does not have mode. Also, a set of data can have more than one mode. When there are two modes, the distribution of values is referred to as bimodal.

More questions? See questions #7, #14, and #16.

What Is an Example of How the Mode Can Be Used?

The mode is the least precise measure of central tendency because it deals with the frequency of a value's occurrence rather than the value itself. However, if one were charged with computing the most central value to represent a set of labels, the mode would be the most appropriate measure of central tendency.

For example, here is a list of the number of members who identify with five different political parties.

Party	Number of Members
A	587
B	456
C	454
D	876
E	194

In this example, the mode is Party D, the party that occurs most frequently.

In the following bimodal example, there are two values that occur with equal frequency.

Party	Number of Members
A	876
B	456
C	454
D	876
E	194

Both Party A and Party D occur with the same frequency and so are both modes in this distribution of scores. Consequently, the distribution is bimodal in nature. A bimodal distribution would have two "high points" or humps, as you see in Figure 14.1.

Figure 14.1 A bimodal distribution

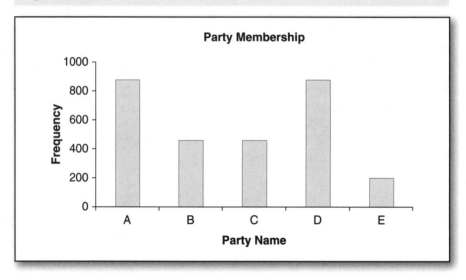

More questions? See questions #7, #13, and #16.

How Can I Decide Which
Measure of Central Tendency to Use?

The specific measure of central tendency you should use depends primarily on the type of data that you are examining.

Here are some summary statements about what measure to chose when.

1. If your data are categorical or nominal in nature, then use the mode. Categorical or nominal data are information that can fit in only one category at a time and whose values can be distinguished only by their labels or names. They hold no quantitative value. Some examples of these are hair color, political affiliation, car model, and favorite baseball team.

For example, at a meeting of political junkies, the values associated with party affiliation might be Republican, Democrat, and Independent. Say one were charged with determining the one most representative value for party affiliation. If the largest number of attendees identify themselves as Republican, that would be the mode and the most accurate measure of central tendency.

2. If data contain extreme scores, the most appropriate measure of central tendency is the median.

For example, if one needed to determine the average house value in a large city, knowing that there are some very expensive homes and some very inexpensive ones, the best and most accurate measure would be the median because it is insensitive to extreme scores.

3. If data are not categorical and do not contain extreme scores, the mean is the most appropriate measure of central tendency to use.

For example, if you needed to compute the average spelling score for a group of 6th graders and the score was defined as the number of words spelled correctly out of 20, the mean would be the correct choice.

Here are two important things to remember about the distinctions among these three measures:

1. The mean is more precise than the median, which is more precise than the mode. When possible, use the mean, the most precise measure.

2. If an outcome can be measured using the mean, then it can probably be measured using the median or the mode as well. Similarly, if an outcome can be measured using the median, it can also be measured using the mode.

More questions? See questions #9, #11, and #13.

How Can Excel Be Used to Compute Measures of Central Tendency?

Excel can be used to compute measures of central tendency by follow-ing these steps. The data represent 50 scores on an achievement test ranging from 70 to 100 (out of a possible 100).

1. Click the Data tab and then click the Data Analysis option.

2. Double-click on the Descriptive Statistics option in the Data Analysis dialog box.

3. Define the Input Range and the Output Range. Also click Labels in first row and the Summary statistics box, as you see in Figure 16.1.

Figure 16.1 The Descriptive Statistics dialog box

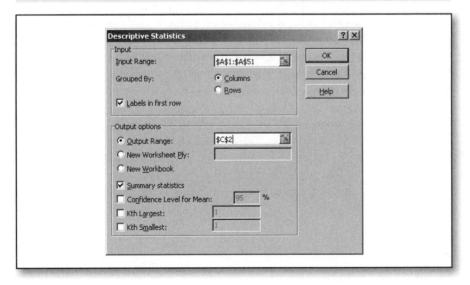

4. Click OK and you will see results of the analysis, as shown in Figure 16.2.

Figure 16.2 The completed descriptive analysis

	A	B	C	D
1	Score			
2	96		*Score*	
3	83			
4	96		Mean	84.46
5	100		Standard Error	1.30
6	80		Median	82
7	77		Mode	97
8	96		Standard Deviation	9.17
9	70		Sample Variance	84.09
10	82		Kurtosis	-1.41
11	97		Skewness	0.19
12	79		Range	30
13	77		Minimum	70
14	91		Maximum	100
15	79		Sum	4223
16	74		Count	50

As you can see, Excel provides a wealth of descriptive information. The mean is 84.46, the median is 82, and the mode is 97.

You can also use Excel to compute the mean, median, and mode by using what are called *functions*. A function is a predesigned formula for computing a value.

To compute the mean, you would use the =AVERAGE function. To compute the median, you would use the =MEDIAN function. And, to compute the mode, you would use the =MODE function.

To use a function, enter the function's name in any cell and then enter the cell range that contains the data of interest.

For example, to compute the mode of the data used in this example, in any cell enter the function =MODE(A2:A51). Once you press Enter, the value of the mode (97) will be returned.

More questions? See questions #9, #11, and #13.

PART 3

UNDERSTANDING MEASURES OF VARIABILITY

What Are Measures of Variability, and Why Are They Used?

Measures of variability such as the range, standard deviation, and variance are the second type of descriptive statistic that one needs to fully describe the characteristics of a set of data. As you already know, measures of central tendency are the first type.

Measures of variability, also called spread or dispersion, reflect how different scores are from one another. More accurately stated, variability is the distance or the amount that any score in a set of scores differs from a particular score. That score, in almost all measures of variability, turns out to be the average score—in particular, the mean.

The three descriptive statistics that are used to measure variability in a set of scores are the range, the standard deviation, and the variance.

The range, which is the least precise of these three measures, is the distance between the lowest and highest value in a set of scores. The range is the easiest value to compute and provides the most general and gross estimate of how much scores differ from one another.

The standard deviation is the average amount that each score differs from a particular score, namely the average or mean for that set of scores. Standard deviation is the most often reported measure of variability. It is also used for many more advanced statistical procedures, which you will read about in later portions of *100 Questions (and Answers) About Statistics*.

The variance is the square of the standard deviation. Like the standard deviation, the variance is used as a descriptive tool as well as in the computation of more advanced statistics.

While each of these measures of variability provides important information for understanding the characteristics of a set of data, variability as a concept is very important to the entire study of statistics. This is because many of the foundational ideas behind techniques, including inferential statistics, use variability as a term that reflects "error." Error should not be confused with mistakes or inaccuracy. Rather, it is the naturally occurring incidence of individual differences occurring between participants in any research and the need for those differences to be captured and understood. That's the role of variability, both in theory and in practice

More questions? See questions #19, #21, and #25.

What Is an Example of How a Measure of Variability Can Be Used?

Variability as a construct is a fascinating topic because it provides insight into between-individual and within-individual changes that occur in human behavior. This is often referred to as *individual differences*. When this idea is applied to topics of current interest, such as child obesity, variability can take on even more meaning as an important measure of change.

Child obesity is frequently associated with dysfunction of the nervous system, in particular the autonomic nervous system or the one over which individuals have no control (it controls breathing and heart rate). These scientists studied whether children undergoing puberty (preteens and teens) were vulnerable to autonomic nervous system problems, such as decrease of heart rate variability, due to problematic metabolic control associated with obesity.

The study by Chen and colleagues explored the influence of pubertal development on autonomic nervous system function in 84 overweight and obese children and 87 nonobese children. The children's autonomic nervous system was studied by measuring heart rate variability. Results of the study showed that the overweight or obese children had significantly lower heart rate variability, and this was positively correlated with their physical activity levels. The researchers concluded that obesity adversely affects the autonomic nervous system function of children, especially during puberty, and that obese children should be encouraged to engage in physical activities during puberty to improve their autonomic nervous system function, which may in turn mitigate obesity.

From an applied perspective, the authors suggested that school nurses should be aware of signs of autonomic nervous dysfunction such as fatigue, agitation, and lethargy, especially for pubescent children. The authors also suggested that school nurses might collaborate with physical education teachers to design appropriate physical activity programs and provide a healthy environment suitable for students' daily activities.

Here's the complete reference . . .

Chen, S., Chiu, H., Lee, Y., Sheen, T., & Jeng, C. (2012). Impact of pubertal development and physical activity on heart rate variability in overweight and obese children in Taiwan. *The Journal of School Nursing, 28*(4), 284–290.

More questions? See questions #17, #22, and #25.

What Is the Range, and How Is It Computed?

The range, which is the most gross measure of variability, is computed by subtracting the lowest score in a set of data from the highest score. It simply provides a measure of the distance between two points without taking into account any of the other points between those two (which is why it is a very general and often imprecise measure). It is, however, a useful tool for getting an idea of the extent of the variability within a set of scores.

There are two kinds of ranges.

The first type is the exclusive range, which is computed as

$$r = h - l,$$

where

r = range,

h = the highest score in the set of scores, and

l = the lowest score in the set of scores.

For example, if a set of scores on an achievement test consists of a low of 47 and a high of 88, then the exclusive range would be

$$r = 88 - 47 = 41$$

The second type is the inclusive range, which is computed as

$$r = h - l + 1$$

It is computed the same way as the exclusive range but with the addition of 1. While the exclusive range is the most commonly computed and presented in research reports, the inclusive range is also an option. Generally, it does not matter which type of range is reported as long as the same type is used consistently throughout a report.

The range is useful in that it provides an overall estimate of how much scores may differ, but since it does not consider any individual scores, it is very limited in its utility.

More questions? See questions #17, #18, and #25.

What Is an Example of
How the Range Can Be Used?

While the range is simple to compute and understand, what the range represents takes on different meanings and importance depending upon the context in which this measure of variability is used.

Three researchers publishing in *Gifted Child Quarterly* examined the range of reading fluency and comprehension scores of 1,149 students in 5 diverse elementary schools, including a school identified as a gifted-and-talented magnet school. Their results show how valuable a measure like the range can be.

The results revealed a range in reading comprehension across all schools of 9.2 grade levels in Grade 3, 11.3 in Grade 4, and 11.6 in Grade 5. These researchers also found a wide range of oral reading fluency scores across all elementary schools, as students scored from below the 10th percentile to above the 90th percentile. Here's where the range is an important piece of information, in that these findings demonstrate the wide range of reading achievement in diverse populations of students, including gifted students. The researchers concluded that teachers need to differentiate reading content and instruction to fit a wide variability in skills to ensure students' success.

Here's the complete reference . . .

Firmender, J. M., Reis, S. M., & Sweeny, S. M. (2013). Reading comprehension and fluency levels ranges across diverse classrooms: The need for differentiated reading instruction and content. *Gifted Child Quarterly, 57*(1), 3–14.

More questions? See questions #17, #18, and #19.

What Is the Standard Deviation, and How Is It Computed?

The standard deviation (represented by a lowercase s) is a measure of how much, on average, each score in a set of scores varies from the average (usually the mean) of that set of scores. One of several measures of variability, it is used to assess how much variability or diversity there is in any one set of scores.

It is computed by finding the average amount that each score deviates from the average of all the scores in the data set.

The formula is

$$s = \sqrt{\frac{\Sigma(X - \bar{X})^2}{n - 1}}$$

Here's an example using a very simple set of scores, which represent the number of correct words on a 20-item spelling test.

16

14

10

15

14

12

19

15

8

7

And, here are the steps for computing the standard deviation:

1. List each of the scores, as above.

2. Compute the average for all 10 scores, which in this case is 13.

3. Subtract each individual score from the average. For example, 16 – 13 is 3.

4. Square each of the deviations. For example, 3 squared, or 3^2, is 9.

5. Compute the sum of all these squared deviations from the average, which is 126.

6. Divide this sum by the number of observations minus 1. Here we are dividing by 10 – 1 or 9. Dividing 126 by 9 gives 14.

7. Compute the square root of 14, which is 3.74. That's the standard deviation.

Extras:

- When calculating the standard deviation, one squares the deviations about the mean because if they were not squared (which makes each value positive), their sum would equal zero.
- The square root is used as a last step in the process to return the values to their original units (before they were squared).
- Each standard deviation represents a certain distance from the average or mean of the set of scores.
- The variance (s^2), another measure of variability, is equal to the square of the standard deviation.

More questions? See questions #17, #19, and #23.

Why Is the Unbiased Estimate of $n - 1$ Used, Rather Than Just the Biased n, in the Computation of the Standard Deviation and the Variance?

You'll remember from our discussion of the computation of the standard deviation, in question #21, that the standard deviation is the average distance of each score from some central point in a set of scores. And, you'll remember that this "average" is computed by dividing the total distance by the number of observations.

But, in this case, instead of using n to represent the size of the sample, we use $n - 1$. Why?

A general idea that drives the use of statistics is that it produces values and results that can be trusted, in the sense that statistical outcomes are as accurate an indicator as possible of what is being examined. Recall that we are using sample values of the standard deviation and the variance to estimate characteristics of a population.

In the case of the standard deviation and the variance, these descriptive measures are thought to be biased in that they may overestimate the true value of the population measures. To compensate for this bias, the denominator in the formula for the standard deviation (and the variance) is reduced by 1, resulting in what is called an unbiased value. Because of this, the computed value may underestimate the true population value, but it is thought to be a more accurate estimate than is the biased value.

And what's so interesting about this correction for bias is that the larger the sample, the more significant the correction from biased to unbiased.

Here's the biased formula for the computation of a standard deviation with a sample size of 10:

$$s = \sqrt{\frac{200}{10}} = 4.47$$

Look what happens when the amount of "variability" (in the numerator) remains constant and the sample size is increased to 100.

$$s = \sqrt{\frac{200}{100}} = 1.14$$

Quite a dramatic change. Why? Because as the sample size gets larger, it more closely approximates the size of the population and is a more accurate estimation (and, of course, the standard deviation is reduced in value). For the most part, papers and journal articles report unbiased estimates.

More questions? See questions #17, #21, and #23.

What Is an Example of How the Standard Deviation Can Be Used?

You are probably not surprised to learn that in most cases, before a medication or drug can be made widely available, it undergoes extensive testing. As a company decides how to allocate resources to a drug's development, it considers many factors such as the complexity of the drug, the need for it, development costs, and competition. One factor is how global the drug's development efforts should be: Can the drug be tested on one regional or national group and those results be generalized to others? This is a big and important question.

When multicountry clinical trials are involved, it is vital to understand the characteristics of each country and the variability of the data reported, as measured by such indicators as the coefficient of variation and the standard deviation. Data variability is one of the most important factors in the precise estimation of treatment effects as well as calculation of sample size when planning the study.

Researchers from Kitasato University in Tokyo compared data from clinical trials in Japan with data from outside Japan on 29 drugs. The authors found that Japanese data are similar to non-Japanese data in terms of data variability and that Japanese and non-Japanese values for the measures of variability were relatively close. They therefore concluded that Japanese clinical trial data showed variability similar to that of non-Japanese data for most cases, meaning that the drugs' effects may be similar across national groups.

Here's the complete reference . . .

Kanmuri, K., & Narukawa, M. (2013). Investigation of characteristics of Japanese clinical trials in terms of data variability. *Therapeutic Innovation & Regulatory Science, 47*(4), 430–437.

More questions? See questions #17, #21, and #24.

What Is an Example of How the Variance Can Be Used?

Domestic or family violence continues to be a significant problem, having long-lasting and significant consequences for survivors. It's a major public health problem across the socioeconomic spectrum as well as a challenge for law enforcement and judicial officials.

However, research on this topic has almost always focused on a patriarchal model, where men are often inaccurately shown as perpetrators and women as victims. Empirical research on sibling abuse in families has been significantly absent from the professional literature. In this study, we have an example of how the variance standing alone is not as useful as the variance as part of a sophisticated analytical technique. As we mentioned earlier, the standard deviation is the most common descriptive statistic used to assess variability. The variance, the square of the standard deviation, is most commonly used (as is the case here) as a component in a much more ambitious analysis.

The research reported by Morrill and Bachman used a survey instrument based on the Conflict Tactics Scale (CTS) to measure whether significant gender differences exist in the experience of sibling abuse as a child, either as perpetrator or victim. MANOVAs (multivariate analyses of variance) indicated that there are no gender differences related to surviving sibling abuse or perpetrating emotional and physical abuse, whereas it was found that women have a significantly higher rate of perpetration related to sibling sexual abuse.

Here's the complete reference . . .

Morrill, M., & Bachman, C. (2013). Confronting the gender myth: An exploration of variance in male versus female experience with sibling abuse. *Journal of Interpersonal Violence, 28*(8), 1693–1708.

More questions? See questions #17, #22, and #25.

How Can Excel Be Used to Compute Measures of Variability?

Both SPSS and Excel can easily be used to compute the range, standard deviation, and variance.

To use Excel to compute these measures of variability you can use either the software's individual functions or the Data Analysis ToolPak. We assume that you have already entered some data as a column of values.

To use the functions, follow these steps:

1. To compute the standard deviation, enter one of the following functions in any free cell (although entering it right below the data is most useful):

   =STDEV.P to compute the population's standard deviation

   =STDEV.S to compute the sample's standard deviation

2. Enter the range of the values for which you want to compute the standard deviation and press Enter.

3. To compute the variance, enter one of the following functions in any empty cell (although entering it right below the data is most useful):

   =VAR.S to compute the sample's variance

   =VAR.P to compute the population's variance

4. Enter the range of the values for which you want to compute the variance and press Enter.

Instead of using these functions embedded in Excel, you can use the Descriptive Analysis option in Excel's Analysis ToolPak to compute all the measures of variability. The ToolPak is only available for the Windows version of Excel.

More questions? See questions #17, #79, and #80.

PART 4

ILLUSTRATING DATA

Is a Picture Really Worth a Thousand Words? Why Illustrate Data?

Before this answer seeks to enlighten with words, take a look at a certain set of scores (number of boxes of cookies sold by week) and a graph of the same.

Here are the data organized as a table.

Week	Boxes Sold
Week 1	12
Week 2	15
Week 3	8
Week 4	22

And Figure 26.1 shows a simple line chart created using Excel.

Figure 26.1 A simple line chart

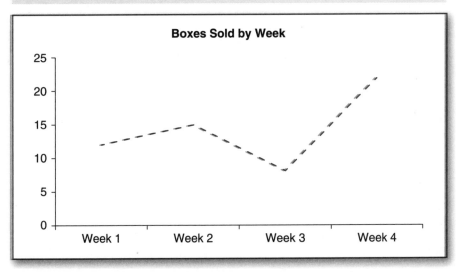

We used a line chart because it is the best format for recording and visualizing change over time. You'll learn how to create a simple line chart like this with just a few mouse clicks later on in this section of *100 Questions* (see question #34).

The important question when comparing these two ways of representing data (the table and the chart) is, Which is easier to understand?

One of our most powerful, and useful, senses is sight. By our nature, we force ourselves to make sense out of visual stimuli. While the table above is compact and direct, by itself, it does not gives you sense of how the data may change over time (weeks in this case) and the magnitude of that change. The chart does both, with the line being either flat or going up or down by some magnitude from one point to another.

For example, the difference between weeks 3 and 4, where there was an increase of 14 boxes sold, becomes especially meaningful when you look at the steep slope of the line between weeks 3 and 4 as compared to, for example, between weeks 1 and 2.

A visual representation of data allows us to see those changes quickly and easily. Always report summaries of data in tables and such but, if possible, chart them as well.

More questions? See questions #27, #34, and #39.

What Are Some Guidelines for Creating Effective Illustrations of Data?

There are hundreds of ways in which you can chart data to make it visually interesting yet not overwhelming. And while we will deal with charts like line charts, pie charts, and bar charts in later sections of *100 Questions (and Answers)*, here are five things to keep in mind regardless of the type of chart being created and the type of data being presented.

1. *Plan, plan, and then plan some more.* Get out some graph paper and actually draw out how you would like the chart to appear, including title, axes titles if necessary, patterns, size, and any other important element of the presentation.

2. *Present one idea per chart.* While you could present several ideas in one chart, focus on just one. The purpose of the chart will be clearer, and the likelihood that a reader will misunderstand the chart will be greatly reduced.

3. *Make sure that the scales (the x- and y-axes) are in proportion to one another.* You want the vertical and horizontal axes to appear in a correct ratio (about 3:4) so that the chart does not look artificially contrived.

4. *Simple is best.* Your goal should be to present a chart that maximizes understanding and minimizes clutter and potential for misunderstanding. The chart should be simple—but not overly so—and convey one major idea. If you need to present additional information that will create visual clutter, save it for a footnote or accompanying text.

5. *No junk.* Chart junk is created when you use every available bell and whistle that the software program offers—including many different designs, patterns, shapes, and sizes—to create a chart. Using all these features may be fun, but they communicate a little bit of everything and nothing of anything. Be conservative: Use as few tools as necessary to get the message across.

More questions? See questions #26, #38, and #39.

What Are Frequency Distribution and Cumulative Frequency Distribution, and How Do I Create Them?

A frequency distribution is a summary of raw data that indicates how often each data point occurs. It is a method for organizing a set of raw data into class intervals, which can then be used to create a histogram. A histogram is a visual representation of a frequency distribution.

Here are 25 scores followed by the steps necessary to create a frequency distribution.

20	24	11	10	2
1	15	23	13	11
1	4	20	13	1
4	13	23	17	14
1	3	5	4	3

1. Select a class interval or a range of scores into which each data point can be placed. Try to have 5–10 class intervals. In this example, we'll use a class interval of five values so the intervals are 0–4, 5–9, 10–14, 15–19, and 20–24.

2. Now, using a two-column table, count the number of times each value appears in the raw data and enter that number next to the appropriate class interval. Your final tally, a frequency distribution, should look something like this:

Class Interval	Frequency
20–24	5
15–19	2
10–14	7
5–9	1
0–4	10

In a *cumulative* frequency distribution, one frequency value is added to the previous one, as shown below, and the cumulative sum of the frequencies is shown. Cumulative frequency distributions are often used along with cumulative percentages to see how much of any one class of scores occupies the entire set of scores.

Class Interval	Frequency	Cumulative Frequency
20–24	5	25
15–19	2	20
10–14	7	18
5–9	1	11
0–4	10	10

Frequency distributions are not a type of chart, as we generally know charts to be more visual in nature, but they are one step above a simple listing of raw data. Furthermore, they are the first step in the creation of a histogram or other charts.

More questions? See questions #26, #29, and #31.

What Is a Histogram, and How Can I Create One Manually?

A histogram is a visual representation of a frequency distribution. It is a simple way to effectively illustrate the number of values that fall into each class interval. While we will show you how to create a histogram using Excel (question #30), it is important to know how to create one manually.

To create a histogram like the one you see below (using the data from question #28) follow the steps on page 57.

Figure 29.1   A simple histogram

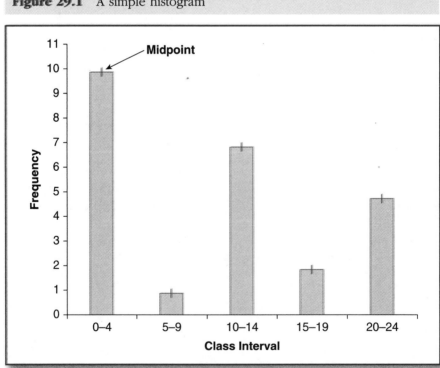

1. Using graph paper with ¼- to ½-inch boxes (you can create your own at www.printfreegraphpaper.com), label the vertical or *y*-axis as "Frequency" and the *x*-axis as "Class Interval." In this example (see the answer to question #28), the *y*-axis can be labeled from 0 to 12 and the *x*-axis as 0–4, 5–9, 10–14, 15–19, and 20–24.

2. Create a column for each of the midpoints of the class interval. The midpoint is the center of the class interval. The height of each bar in the histogram corresponds to the frequency of that class interval. For example, for the class interval of 10–14, which has a midpoint of 12, the height of the bar representing frequency should be at the value of 7.

   a. You can also make tally marks for each occurrence of a score, resulting in the same visual representation but no bars—just a set of tally marks stacked upon one another.

The value of a histogram is that you can easily get a sense of what class of values occurs more frequently than others, get a rough estimate of the range of scores, and get an idea of the average (such as the median or the mean).

More questions? See questions #26, #30, and #39.

How Can I Create a Histogram Using Excel?

This option only works if you are using Excel for the PC, not for a Mac (as of Mac version 2011, which does not contain the Data Analysis ToolPak). If you want to create a histogram using a Mac, use Excel's Column Chart option.

To create a histogram using Excel and the data from question #28, follow these steps:

1. Convert the class intervals (0–4, 5–9, 10–14, 15–19, 20–24) to 1, 2, 3, 4, and 5, respectively. Place one score entry in the Excel worksheet for each occurrence in the set of scores, such as you see in Figure 30.1.

2. Create a column and enter the "bins," which are numbers that represent each of categories found in the Score column (1–5 in this example), as shown in Figure 30.1.

Figure 30.1 Entering data for a histogram

	A	B
1	Score	Bin
2	1	1
3	1	2
4	1	3
5	1	4
6	1	5
7	2	
8	2	

3. Click Data Tab → Data Analysis and then select Histogram from the Data Analysis dialog box.

4. Enter the range of cells for the Input Range, the range of cells for the Bin Range, and the range of cells for the Output Range.

5. Click the Labels box.

6. Click the Cumulative Percentage and the Chart Output boxes, and you will see the completed Histogram dialog box as shown in Figure 30.2.

Figure 30.2 Histogram dialog box

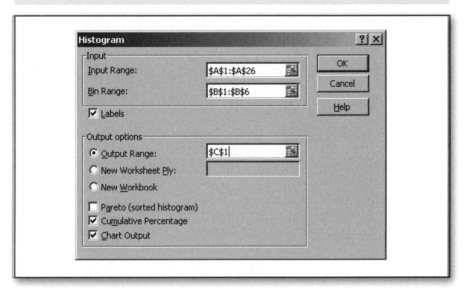

7. Click OK, and you will see the histogram as shown in Figure 30.3.

Figure 30.3 The completed histogram

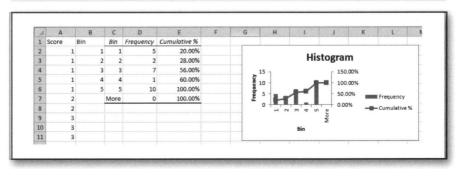

The value of a histogram is that you can easily get a sense of what class of values occurs more frequently than others, get a rough estimate of the range of scores, and get an idea of the average (such as the median or the mean).

More questions? See questions #29, #32, and #39.

What Is an Ogive, and How Can I Create One Using Excel?

A histogram is a visual representation of a frequency distribution. An ogive is a visual representation of a cumulative frequency distribution. It is very useful because it shows the change in the number of scores within each class interval.

To create an ogive using Excel, follow these steps:

1. In a new worksheet, enter the class interval and the number of times a score occurs within that class, as was done in question #30. This is a frequency distribution on which the creation of the ogive will be based. Be sure to label the columns "Class Interval" and "Frequency."

2. Create two new columns. New column 2 will be the midpoint of each class interval. New column 4 will be where the cumulative frequencies are entered. The cumulative frequency is the number of scores in the class plus the sum of the number of scores below that class. All the columns are shown in the following table.

Class Interval	Midpoint	Frequency	Cumulative Frequency
20–24	22	5	25
15–19	17	2	20
10–14	12	7	18
5–9	7	1	11
0–4	2	10	10

3. Highlight the second (Midpoint) and the fourth (Cumulative Frequency) columns simultaneously.

If you are using the PC version of Excel, follow these steps:

4. Click the Insert tab and then click the Scatter icon.

5. From the Scatter icon drop-down menu, click the Scatter with Smooth Lines option. From the drop-down choices on the Scatter menu, click the Smooth Lined Scatter option. You will see an ogive like the one shown in Figure 31.1

Figure 31.1 An ogive or a cumulative frequency distribution

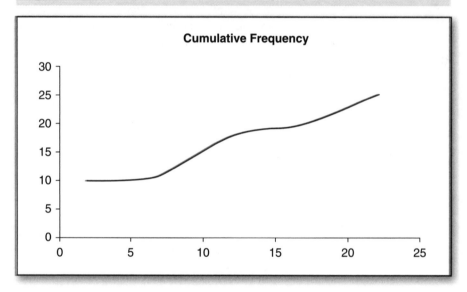

If you are using the Mac version of Excel, follow steps 1 through 3 above and then do the following:

4. Click the Charts tab and then click the Scatter chart icon.

5. From the Scatter icon drop-down menu, click the Smooth Lined Scatter option.

Note that if you use the Histogram tool in the PC version of Excel, you can create an ogive by clicking on the Chart option in the Histogram dialog box.

Whether you use the Mac or the PC version of Excel, be sure that the cells containing the class intervals are formatted as Text so that Excel knows to use them as axes labels and not as numerical values.

More questions? See questions #28, #30, and #32.

What Is a Column Chart, and How Can I Create One Using Excel?

A column chart is a visual representation of categorical data that uses vertically oriented bars to represent frequencies or values of the x-axis. To create a column chart using Excel, follow these steps:

1. In a new worksheet, enter the categories as well as the number of times each category appears, as shown in the following table.

2. Highlight all the data.

Party	Number
Democrat	154
Republican	213
Independent	54

If you are using the PC version, do the following:

3. Click the Insert tab and then click the Column icon.

4. Click the 2-D Column icon to produce the chart you see in Figure 32.1.

If you are using the Mac version of Excel, follow steps 1 and 2 above and then

3. Click the Charts tab and then click the Column icon.

4. Click the Clustered Column icon. You will see the chart as shown in Figure 32.1.

Figure 32.1 A simple column chart

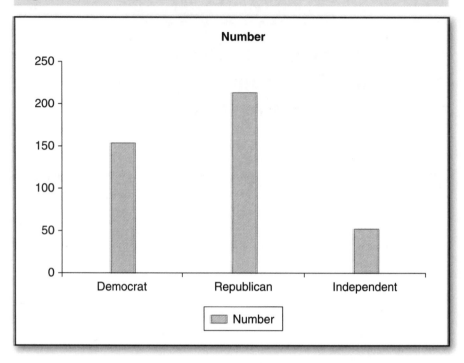

Whether you use the Mac or the PC version of Excel, be sure that the cells containing the category names are formatted as Text so that Excel knows to use them as axes labels and not as numerical values.

More questions? See questions #29, #30, and #39.

What Is a Bar Chart, and How Can I Create One Using Excel?

A bar chart is a visual representation of categorical data that uses horizontally oriented bars to represent frequencies or values of the *x*-axis. To create a simple bar chart using Excel, follow these steps:

1. In a new worksheet, enter the data as shown here. In this case, gender and frequency are represented.

Gender	Frequency
Male	156
Female	210

2. Highlight all the data.

If you are using the PC version, do the following:

3. Click the Insert tab and then click the Bar icon.

4. Click the 2-D Bar icon to produce the bar chart you see in Figure 33.1.

If you are using the Mac version of Excel, follow steps 1 and 2 above and then

5. Click the Charts tab and then click the Bar chart icon.

6. Click the Clustered Bar icon. You will see the chart as shown in Figure 33.1.

Figure 33.1 A simple bar chart

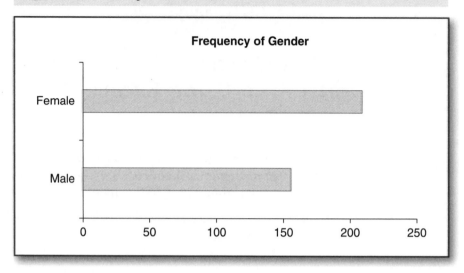

Whether you use the Mac or the PC version of Excel, be sure that the cells containing the values for the *x*-axis are formatted as Text so that Excel knows to use them as axes labels and not as numerical values.

More questions? See questions #26, #32, and #39.

What Is a Line Chart, and How Can I Create One Using Excel?

A line chart is a visual representation of noncategorical data that uses a line to represent values of the *x*-axis. To create a line chart using Excel, follow these steps:

1. In a new worksheet, enter the categories as well as the number of times each category appears as in the following table, which shows quarterly income in thousands of dollars.

Quarter	Income
Quarter 1	$1,867
Quarter 2	$2,193
Quarter 3	$989
Quarter 4	$1,358

2. Highlight all the data.

If you are using the PC version, do the following:

3. Click the Insert tab and then click the Line icon.

4. Click the 2-D Line icon to produce the chart you see in Figure 34.1.

Figure 34.1 A simple line chart

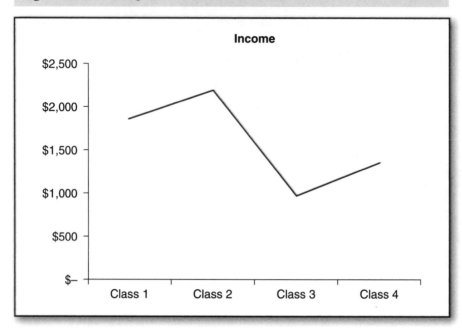

If you are using the Mac version of Excel, follow steps 1 and 2 above and then

3. Click the Charts tab and then click the Line icon.

4. Click the 2-D Line icon. You will see the chart as shown in Figure 34.1.

More questions? See questions #37, #38, and #39.

What Is a Pie Chart, and How Can I Create One Using Excel?

A pie chart is a visual representation of categorical data that uses sections within a circle (called slices) to represent frequencies of the different levels of a variable. To create a pie chart using Excel, follow these steps:

1. In a new worksheet, enter the categories as well as the number of times each category appears, as shown in the following table. Here we have data on the number sold of four models of automobile:

Model	Number Sold
Volvo	564
Chevrolet	3,434
Honda	4,331
Mercedes	312

2. Highlight all the data.

If you are using the PC version of Excel:

3. Click the Insert tab and then click the Pie icon.

4. Click the first 2-D Pie icon to produce the chart you see in Figure 35.1.

Figure 35.1 A simple pie chart

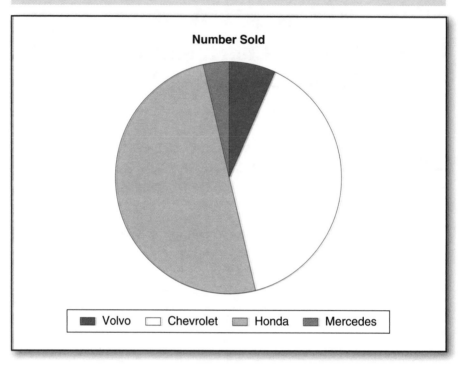

If you are using the Mac version of Excel, follow steps 1 and 2 above and then do the following:

3. Click the Charts tab and then click the Pie icon under 2-D Pie. You will see the chart as shown in Figure 35.1.

More questions? See questions #26, #37, and #38.

What Is a Scatter Chart, and How Can I Create One Using Excel?

A scatter chart is a visual representation of two data points for each case in the data set, for example, the height and weight for each of 10 participants in a weight loss program. To create a scatter chart using Excel, follow these steps:

1. In a new worksheet, enter two scores for each case (both height and weight), as shown in the following table.

Participant	Height (inches)	Weight (pounds)
1	66	154
2	54	123
3	71	214
4	65	265
5	74	276
6	61	205
7	52	116
8	73	245
9	78	235
10	59	156

2. Highlight all the data.

If you are using the PC version of Excel, do the following:

3. Click the Insert tab and then click the Scatter icon.

4. From the Scatter drop-down menu, click the first Scatter icon to produce the chart you see in Figure 36.1. Here you can see how

Figure 36.1 A simple scatter chart

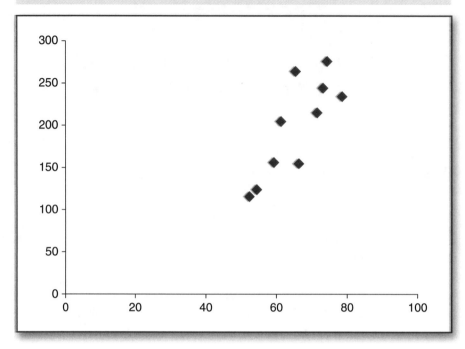

each data point in the chart represents two scores, one from the *y*-axis and one from the *x*-axis.

If you are using the Mac version of Excel, follow steps 1 and 2 above and then

3. Click the Charts tab and then click the Scatter icon.

4. From the Scatter drop-down menu, click Marked Scatter. You will see the chart shown in Figure 36.1.

More questions? See questions #38, #42, and #43.

QUESTION #37

How Can I Edit Any
Chart I Create in Excel?

There are almost an unlimited number of ways to edit any chart in Excel. In Figure 37.1, you can see an edited version of the scatter chart that appeared originally in Figure 36.1.

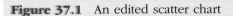

Figure 37.1 An edited scatter chart

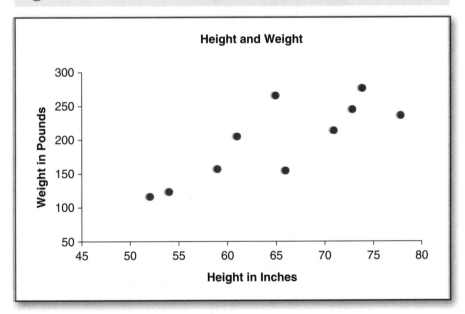

To edit this chart, we did the following:

1. Clicked the Chart Area of the chart, which appears as you move the mouse over the chart.

2. Clicked any one of the Chart Layout options appearing in the Excel ribbon.

3. Clicked the Chart Title area and edited it to read "Height and Weight."

4. Clicked the vertical Axis Title and edited that to read "Weight in Pounds."

5. Clicked the horizontal Axis Title and edited that to read "Height in Inches."

6. Clicked the Series1 legend and pressed the Delete key to delete it.

7. Double-clicked on the diamond-shaped marker (◆) in the chart to reveal the Format Data Series dialog box, as you see in Figure 37.2.

Figure 37.2 The Format Data Series dialog box

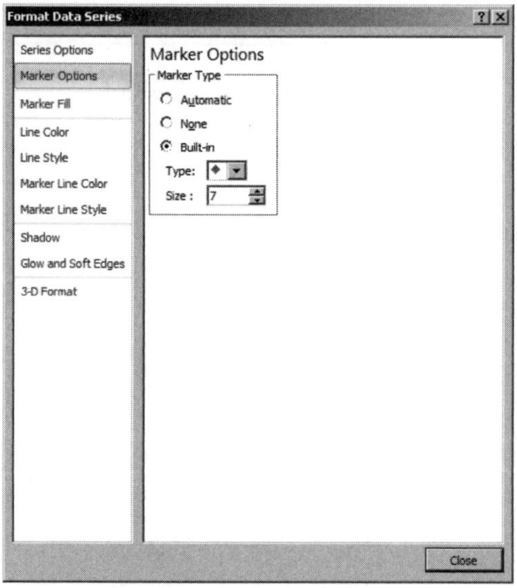

8. Clicked Marker Options and Built-in and selected the solid circle from the Type drop-down menu.

9. Clicked Close.

The most important thing to know about editing any chart is that you can double-click on any element of that chart (axis, markers, lines, etc.) and edit as you see fit.

More questions? See questions #26, #36, and #39.

QUESTION #38

How Can I Integrate a Chart Into Other Documents?

When working with any set of data, you should only have to enter that data once in a digital format. For example, if you create a chart using a spreadsheet such as Excel, you should be able to easily import that chart into a document created with Word or OpenOffice Writer or some other word processor.

To import a chart to another document, follow these steps:

1. Highlight the chart, as you see in Figure 38.1, by clicking once anywhere in the chart area. When selected, the border of the chart changes from being a thin black line to a wider gray edge.

Figure 38.1 Highlighting a chart in Excel

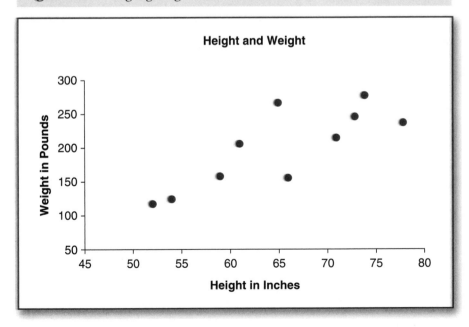

2. Right-click and select Copy or click Copy from the Home tab.

3. Open the document into which you want to paste or integrate the chart and place the cursor where you want the chart to appear.

4. Click Paste from the Home tab and select the Keep Source Formatting & Link to Excel option under Paste Options. You can also right-click and select from the Paste options. This will ensure that any changes in the Excel chart will also appear in the copy of the chart in the Word document.

The chart will appear in the document.

Perhaps the most useful feature when integrating charts created with one application into another is the one that links the copy to the original. If the data in the document where the chart originated (in this case, Excel) are changed, then the copy of the chart will automatically change (in Word, in this example). This feature is usually available in applications that are developed by the same manufacturer—such as Excel and Word, both created by Microsoft. But do keep in mind that if the data are changed, then all files containing the data or charts based on the data have to be saved.

More questions? See questions #26, #37, and #39.

When Should I Use a Chart, and What Type Should I Use?

The answer to the first question is easy—whenever a visual representation of the data you are working with will facilitate a reader's understanding of the idea you are trying to convey. Unless a chart is filled with chart junk, containing far too many elements, almost any visual representation can enhance the presentation of data. Still, while charts are quite helpful, one needs to be judicious in their use: Not every bit of data needs to have an accompanying chart or visual representation. Nonetheless, your most important points will get across more emphatically if they are accompanied by a chart.

As to what chart to use when, here's a brief summary of chart types, when each type should be used, and an example of that use. Note that many different charts can be used to illustrate the same data, but surely some are used more correctly than others.

Chart Type	When to Use It	Example
Column charts	To compare items to one another or examine changes over time. Columns are vertical.	Numbers of pro and anti demonstrators
Bar charts	To compare items to one another or examine changes over time. Bars are horizontal.	Amount of money raised by three community organizations in 2013
Line charts	To display data that are continuous over time. Best used for showing trends in data at equal intervals.	Change in sales each month for one year

Chart Type	When to Use It	Example
Pie charts	To show the size of one item proportionally to the rest of the items	Sales of different types of candy
Area charts	To emphasize the amount of change over time	Growth in one country's population compared to that of nine other developing countries
Scatter charts	To show the relationship between two data points across several cases	Grade point average in high school and grade point average in college
Doughnut charts	Like a pie chart, to show the size of one item proportionally to the rest of the items, but allow for more than one series at a time to be illustrated	Quarterly sales for three years of data

More questions? See questions #27, #37, and #38.

PART 5

UNDERSTANDING RELATIONSHIPS

What Is the Correlation Coefficient, and How Is It Used?

You already know that most collections of data can be described using a measure of central tendency, such as the mean, and a measure of variability, such as the standard deviation. Sometimes, however, it is important to be able to describe the relationship between two or more variables.

The correlation coefficient, also known as the Pearson product-moment correlation coefficient (named after Karl Pearson), is a numerical index that reflects the relationship between two variables such as X and Y.

It is represented by the symbol r_{xy}, and it ranges in value from −1.00 to +1.00. Perhaps the most-frequently made error regarding the correlation coefficient is that absolute value is more significant than the sign. For example, a correlation of −.7 between two variables is stronger than a correlation of +.6.

The correlation between two variables is often referred to as bivariate (for two variables). Correlations between any two variables from a set of several variables can easily be computed. For example, if the variables age, height, and weight are being correlated, one might look at the following relationships:

Variables	Correlation
age and height	$r_{age} \cdot r_{height}$
age and weight	$r_{age} \cdot r_{weight}$
height and weight	$r_{height} \cdot r_{weight}$

The Pearson correlation looks at the relationship between two variables, and both of those variables must be continuous in nature. That is, they can assume any value along a scale of values such as height (inches), weight (pounds), time (in seconds), or income. For example, one might use the Pearson correlation to examine the relationship between years of education and scores on a standardized test. For variables that are not continuous, but are categorical, such as gender or ancestry or voting preference, other correlation techniques can be used.

More questions? See questions #41, #42, and #50.

What Is an Example of How the Correlation Coefficient Can Be Used?

Correlations are always used to make judgments about the strength of the relationship between variables. Sometimes they are used descriptively to see only whether variables are related, but they are also used inferentially, to extend a conclusion to a population based on an earlier finding.

One study that shows how correlation coefficients can be used to infer outcomes from a sample to a population was conducted to examine the relationship between collaborative cultures in schools and student achievement. School culture data were collected from the teachers in 81 schools in Indiana using a survey methodology that was based on six factors.

These six factors were then correlated with student achievement to determine whether features of collaborative cultures (such teachers supporting one another) tend to correlate with higher test scores. All six of these factors—Collaborative Leadership, which describes facilitating collaboration among teachers; Teacher Collaboration, behaviors that are expressive of collaborative cultures; Professional Development, attitudes teachers have toward new ideas; Unity of Purpose, how the school's mission statement influences teaching; Collegial Support, collegiality among teachers; and Learning Partnership, the quality of teacher-parent communications—were positively correlated with student achievement. The findings suggest that these outcomes are associated, so the next step might be giving teachers more formal training to develop such skills and conducting an experimental examination of outcomes in achievement.

Here's the complete reference . . .

Gruenert, S. (2005). Correlations of collaborative school cultures with student achievement. *NASSP Bulletin, 89,* 43–55.

More questions? See questions #40, #44, and #49.

What Are the Different Types of Correlation Coefficients?

The magnitude (size) and the value (ranging from –1.0 to +1.0) gives us lots of information about the relationship between variables and how one variable may change as another does.

If variables change in the same direction, then the correlation coefficient is direct or positive. For example, as children get taller, they generally get heavier as well, meaning that height and weight are directly or positively correlated. This correlation would have a value between .00 and +1.00.

If variables change in opposite directions, then the correlation is indirect or negative. For example, in general the faster one completes a test, the more likely there will be errors. This means that amount of time and error rate are indirectly or negatively correlated. This correlation would have a value between .00 and –1.00.

The following chart provides a summary of how variables can change with regard to each other, the type of correlation that change represents, the value the correlation can assume, and an example.

What Happens to Variable X	What Happens to Variable Y	Type of Correlation	Value	Example
X increases in value.	Y increases in value.	Direct or positive	Ranging from .00 to 1.00	The more hours you spend studying, the higher your test score.
X decreases in value.	Y decreases in value.	Direct or positive	Ranging from .00 to 1.00	The less time you spend sleeping, the worse your test performance.

What Happens to Variable X	What Happens to Variable Y	Type of Correlation	Value	Example
X increases in value.	Y decreases in value.	Indirect or negative	Ranging from .00 to −1.00	The more you exercise, the less you weigh.
X decreases in value.	Y increases in value.	Indirect or negative	Ranging from .00 to −1.00	The less you practice, the more time you have for other activities.

More questions? See questions #40, #41, and #46.

How Can Scatter Charts Help Me Understand Correlation Coefficients?

As you learned earlier, a scatter chart is a visual representation of two data points for each case. Creating a scatter chart is simply a matter of placing coordinates on an *X-Y* grid, such as you see in Figure 43.1 for the following data.

Participant	Height (inches)	Weight (pounds)
1	66	154
2	54	123
3	71	214
4	65	265
5	74	276
6	61	205
7	52	116
8	73	245
9	78	235
10	59	156

Figure 43.1 A scatter chart showing a positive correlation

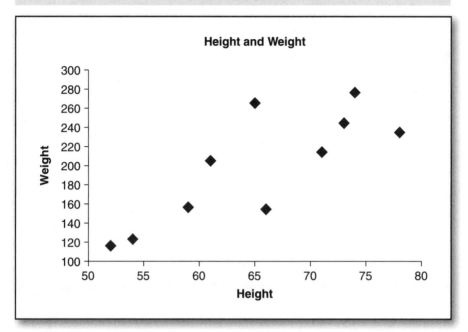

When you look at Figure 43.1, you can see that the data points are organized from the lower left-hand corner of the chart to the upper right-hand corner. This arrangement indicates that the slope of these points is positive as is the correlation between the variables (it is actually .81). Figure 43.1 is a visual representation of a positive correlation. As height increases, weight increases and vice versa.

Figure 43.2 is a visual representation of a negative correlation; the data points tend to group themselves from the upper left-hand corner to the lower right-hand corner of the scatter chart. As time increases, errors decrease and vice versa.

Finally, there are times when variables are unrelated to each other and share nothing in common. This relationship is shown in the scatter chart in Figure 43.3.

Scatter charts are useful because they provide visual clues as to the relationship between variables, thereby illuminating the nature of a relationship.

Figure 43.2 A scatter chart showing a negative correlation

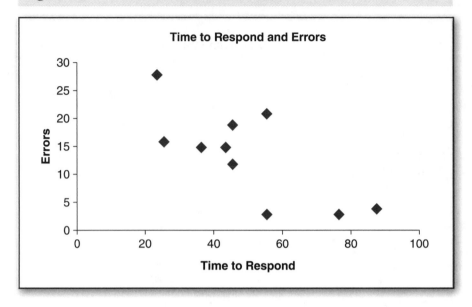

Figure 43.3 A scatter chart showing no correlation

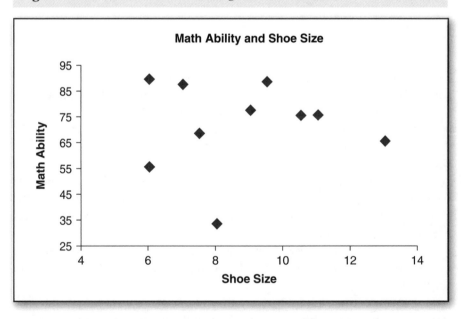

When reporting correlations of interest, it is very useful to include a scatter chart as well.

More questions? See questions #36, #42, and #44.

How Do I Compute a Correlation Coefficient?

A correlation coefficient is computed by inserting the proper values into a simple question.

The formula is as follows:

$$r_{xy} = \frac{n\Sigma XY - \Sigma X \Sigma Y}{\sqrt{[n\Sigma X^2 - (\Sigma X)^2][n\Sigma Y^2 - (\Sigma Y)^2]}},$$

where

r_{xy} is the correlation coefficient between X and Y;

n is the size of the sample;

X is the individual's score on the X variable;

Y is the individual's score on the Y variable;

XY is the product of each X score times its corresponding Y score;

X^2 is the individual's X score, squared; and

Y^2 is the individual's Y score, squared.

For example, let's compute the correlation coefficient, represented as a lowercase r_{xy} for the the variables X and Y, using the following data:

	X	Y	X²	Y²	XY
	1	3	1	9	3
	4	7	16	49	28
	5	8	25	64	40
	3	9	9	81	27
	5	6	25	36	30
	6	9	36	81	54
	7	8	49	64	56
	5	8	25	64	40
	6	9	36	81	54
	4	7	16	49	28
Sum	46	74	238	578	360

With the values substituted, the equation looks like this:

$$r_{xy} = \frac{10(360) - (46)(74)}{\sqrt{[10(238) - (46)^2][10(578) - (74)^2]}} = .69$$

And the result is that r_{xy} equals .69, a positive or direct correlation.

More questions? See questions #40, #45, and #46.

How Can I Use Excel to Compute a Correlation Coefficient?

To compute the correlation coefficient using Excel, you use the =CORREL function. We will be using the same set of data you saw in question #44.

If you are using a Mac or a PC, follow these steps:

1. Enter =CORREL in the cell where you want the results to appear.

2. Enter the first data array, a comma, and then the second data array. The completed function should appear as shown in Figure 45.1.

Figure 45.1   The =CORREL function

	A	B	C	D	E	F
	B12		f_x =CORREL(A2:A11,B2:B11)			
	Q45 data					
1	X	Y				
2	1	3				
3	4	7				
4	5	8				
5	3	9				
6	5	6				
7	6	9				
8	7	8				
9	5	8				
10	6	9				
11	4	7				
12		0.692				

3. Press the Enter key, and as you will see, the correlation of .69 (rounded to the nearest 100th) is returned to the cell in which the function is located.

If you are using the Windows version of Excel, you can also take advantage of the Correlation option in the Data Analysis ToolPak. To do this, follow these steps:

1. Click Data ➔ Data Analysis, and you will see the Data Analysis ToolPak dialog box.

2. Double-click the Correlation option.

3. Complete the Correlation dialog box as you see in Figure 45.2, indicating the Input Range, the Output Range, and whether you want to include column labels in the output.

Figure 45.2 The Correlation Data Analysis ToolPak dialog box

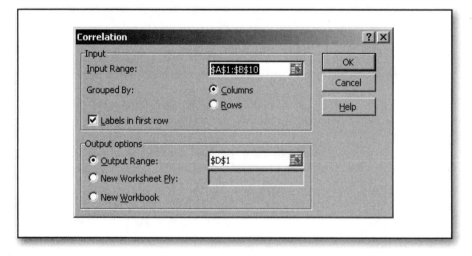

4. Press the Enter key, and the resulting correlation matrix will appear as shown in Figure 45.3.

Figure 45.3 The results of the Correlation ToolPak option

▲	A	B	C	D	E	F
1	X	Y			*X*	*Y*
2	1	3		X	1	
3	4	7		Y	0.689695	1
4	5	8				
5	3	9				
6	5	6				
7	6	9				
8	7	8				
9	5	8				
10	6	9				

More questions? See questions #40, #44, and #46.

What Is the Most Direct Way to Interpret the Value of a Correlation Coefficient?

The correlation coefficient is a numerical value that indicates the degree of relationship between two variables. The value of any correlation coefficient ranges from –1.00 (a perfectly indirect or negative correlation) to +1.00 (a perfectly direct or positive correlation). But how does one interpret this value? What does it mean?

The most direct way to interpret a correlation coefficient is to use the following table. This will give you a quick assessment of the strength of the correlation.

If the absolute value of the correlation coefficient ranges from . . .	The relationship between the variables is . . .
.8 to 1.0	Very strong
.6 to .8	Strong
.4 to .6	Moderate
.2 to .4	Weak
.0 to .2	Absent

While using the above table is not the most precise way of interpreting the strength of a correlation coefficient, it certainly provides a sense of the strength of the relationship between variables. For a more precise method, we'll turn to question #47, which deals with the coefficient of determination.

More questions? See questions #40, #42, and #47.

What Is the Coefficient of Determination, and How Is It Computed?

Although a simple examination of the value of a correlation coefficient can provide a general assessment of the strength of the relationship between two variables, there is a far more precise way to do this.

The coefficient of determination, represented as r_{xy}^2 or the square of the correlation coefficient, yields the amount of variance in one variable that is accounted for by changes in another variable. The concept of the coefficient of determination is based on the fact that variables that are correlated with one another share something in common. The stronger the relationship (r_{xy}), the more they share and the higher the coefficient of determination.

For example, let's take the correlation between height and weight for a class of sixth graders, which is found to be .85. Thus, the simple Pearson product-moment correlation or r_{xy} = .85. The coefficient of determination is .7225, which means that 72.25% of the variance in height (how much sixth graders differ from each other in height) can be accounted for by the variance in weight (how much sixth graders differ from each in weight).

There are several important things to remember about the use of this coefficient.

The first is that the stronger the simple correlation, the more variance is accounted for. For example, if the correlation between one variable and the other is .4, then only .16 or 16% of the variance in one is accounted for by the relationship between variables. If the correlation between one variable and another is .6, then .36 or 36% of the variance is accounted for.

The second (related to the first) is that the more that one variable has in common with a second variable (and the stronger the correlation), the larger the coefficient of determination will be.

If there is no relationship between variables (when r_{xy} = 0), then nothing is shared between the two variables, and no variance or change in one can be accounted for by a change in the other.

More questions? See questions #44, #45, and #46.

What Are Some Important Things to Remember About Understanding and Using Correlation Coefficients?

Several important points are key to understanding and using correlation coefficients in the study of statistics.

1. Correlation coefficients reflect the amount of variability or dispersion that is shared between two variables. The more that is shared, the stronger the relationship. If one variable has no variability and does not differ among the individuals being assessed, then the correlation is zero (the two variables being compared share nothing).

2. The Pearson product-moment correlation coefficient is represented by the small letter r, with a subscript representing the variables that are being correlated. For example,

    r_{xy} is the correlation between variable X and variable Y.

    $r_{strength \cdot speed}$ is the correlation between strength and speed.

    $r_{attitude \cdot outcomes}$ is the correlation between attitude and outcome behaviors.

3. There are two types of correlation coefficients:

    - Direct or positive, where variables change in the same direction
    - Indirect or negative, where variables change in opposite directions

4. The value of the Pearson product-moment correlation can range from −1.00 to +1.00.

5. The value of the coefficient of determination, which is the square of the correlation coefficient, can range from 0% to 100%.

6. The absolute value of the correlation coefficient reflects the strength of the correlation. So a correlation of −.70 is stronger than a correlation of +.50.

7. Indirect or negative correlations, which have a negative sign (such as −.13 or −.87) are neither "better" nor "worse" than direct or positive correlation coefficients—just very different.

8. A correlation always reflects a situation in which there are at least two data points (or variables) per case.

9. A correlation indicates nothing about the causal relationship between variables; it reflects only the strength of their association.

10. Scatter charts are the best choice to display the visual relationship between points whose correlation is being explored.

More questions? See questions #40, #46, and #47.

How Can I Use a Matrix to Display Several Correlation Coefficients?

A single correlation coefficient is always expressed as one value—a numerical index of the strength of the relationship between two, and only two, variables.

If more than two variables are being correlated with one another, then a correlation matrix can help consolidate the information about the data and make the various relationships much easier to understand.

For example, here are some data, and the data table is followed by a matrix that reflects the various correlations that might be computed. Remember that simple Pearson product-moment correlations always reflect the correlation between just two variables.

Here are the data, with each variable named and the possible range in parentheses.

Year in School (1–4)	GPA (0.0–4.0)	Study Hours (0–40)	Boredom Scale (1–100)
1	3.3	23	16
2	3.2	24	26
2	2.8	12	28
3	3.0	12	11
1	1.8	18	57
2	1.8	7	69
3	2.8	22	58
4	2.3	25	44
3	4.0	30	4
4	2.7	9	29

And here is the correlation matrix. These are the correlations between all possible pairs of variables. The number of variables is four, and the possible pairs are four taken two at a time, for a total of six.

	Year in School (1–4)	GPA (0.0–4.0)	Study Hours (0–40)	Boredom Scale (1–100)
Year in School (1–4)	1.00	0.13	0.00	−0.33
GPA (0.0–4.0)		1.00	0.56	−0.86
Study Hours (0–40)			1.00	−0.33
Boredom Scale (1–100)				1.00

As you can see, each cell represents a correlation between two variables. For example, the correlation between grade point average (GPA) and number of hours studying per week is .56, indicating that the more time spent studying, the higher the GPA is and the higher the GPA, the more time is spent studying.

Each of these values represents only the correlation between two variables. The values along the diagonal are 1 because the correlation between one variable and itself is always 1.

Since correlations can be expressed as r_{xy} or r_{yx}, the cells beneath the diagonal of 1s are empty. The correlation between Year in School and score on the Boredom Scale is exactly the same as the correlation between score on the Boredom Scale and Year in School.

More questions? See questions #44, #46, and #47.

What Are Some Other Measures of Correlation, and How Are They Used?

The Pearson product-moment correlation is only one way to compute a correlation, and it is most useful when looking at the relationship between two variables that are continuous in nature, that is, variables that can take on any value along a continuum. One example is the number of words spelled correctly on a spelling test.

What about all those other situations in which variables are not continuous and can only have values at specific points on some scale? These include such categorical variables as gender (male or female or yes or no or 1 or 2) or party affiliation (Republican, Democrat, or other).

The following chart shows what type of correlation to use when you have such sets of variables. In this chart, categorical variables are called nominal, ranked variables are called ordinal, and variables that are continuous are called interval.

Table 50.1 Correlation coefficient shopping, anyone?

Variable X	Variable Y	Type of Correlation	Correlation Being Computed
Nominal variables (voting preference: Republican, Democrat, or other)	Nominal variables (sex: male or female)	Phi coefficient	The correlation between voting preference and sex
Nominal variables (social class: high, medium, or low)	Ordinal variables (rank in high school graduating class)	Rank biserial coefficient	The correlation between social class and rank in high school graduating class

Variable X	Variable Y	Type of Correlation	Correlation Being Computed
Nominal variables (family configuration: intact or single-parent)	Interval variables (grade point average)	Point biserial	The correlation between family configuration and grade point average
Ordinal (height converted to rank)	Ordinal (weight converted to rank)	Spearman rank coefficient	The correlation between height and weight
Interval (number of problems solved)	Interval (age in years)	Pearson correlation coefficient	The correlation between number of problems solved and age in years

Computing these various types of correlations coefficients is beyond the scope of *100 Questions (and Answers) About Statistics*, but there are many books and online resources and computer software packages that can help you do such.

More questions? See questions #41, #46, and #48.

PART 6

UNDERSTANDING MEASUREMENT AND ITS IMPORTANCE

Why Is Measurement an Important Topic for Statisticians to Understand?

When we know about the important ideas that form an introduction to the study of measurement, we have a basic set of tools that we can use to better make sense of all that data we collect, or plan to collect, as part of a research study.

Measurement is the assignment of values or labels to outcomes. Some examples of outcomes include the number of answers that are correct on an achievement test (such as 72), the number of and type of cars sold over a specific weekend (such as 82 Hondas), or the color of a particular paint (such as deep blue). All of these measurements reflect a certain event that informs us within the context of our research question and the hypotheses that are being explored.

There are several reasons why the study of measurement is important within the framework of introductory statistics.

First, all of statistics deals with the measurement of outcomes, so to get started with statistics, one needs to understand the process of measurement and its varied uses.

Second, to measure is to understand the nature of what is being measured. There are many different ways to assess an outcome (such as "62 inches tall" or "taller than his peers"), and the way an outcome is measured relates directly to the questions we are asking, such as "How tall is he?" versus "Is he as tall as his peers?"

Finally, without precise measurement, hypotheses just cannot be accurately tested. In fact, if our measuring tools are faulty, it is difficult to tell whether the hypothesis we are testing is a reasonable explanation for what we observe. Perhaps the way in which that outcome was measured is at fault and the results really cannot be understood, let alone provide answers to the questions being asked.

Accurate measurement involves using tools that are authentic, in that they do what they are supposed to and they are consistent. While the use of such tools does not by itself answer our questions, their use is the first step on the way to that answer.

More questions? See questions #52, #53, and #56.

What Are Levels of Measurement, and Why Are They Important?

L evels of measurement (sometimes called scales of measurement) are the particular levels at which an outcome is assessed. There are four scales of measurement—nominal, ordinal, interval, and ratio—and each has its own set of characteristics.

Nominal levels of measurement are characterized by the categorical nature of the variable being measured. Variables such as political affiliation (party 1, party 2, or party 3), hair color (auburn, black, blond), and gender (male or female) are all categorical variables. This is the least precise level of measurement.

Ordinal levels of measurement are characterized by rank. Variables such as rank in class, order of finishing a race, and best-to-worst school project are ordinal in nature.

Interval levels of measurement are characterized by an underlying continuum that has equally appearing intervals. Variables such as spelling test score (from 1 to 10 correct) and number of pushups are interval-level variables. An important characteristic of the interval level of measurement is that the intervals along the scale on which the variable is based are equal so that, for example, 10 pushups is twice as many as 5.

Finally, ratio levels of measurement are those that have an absolute zero, representing an absence of what's being measured. While in the physical and biological sciences, such variables are common (such as the absence of light or absolute zero temperature or age of a newborn), in the behavioral and social sciences, that's not the case. Rarely is there a total absence of any variable or construct (such as intelligence or aggression).

Following are the most important things to remember about levels of measurement:

- The level of measurement will help determine how results are analyzed.
- Levels of measurement have an order, from nominal being "lowest" to ratio being "highest."
- The higher the level, the more precise the measurement.

- Any one level incorporates all the characteristics of lower levels. For example, you can measure height as Group A and Group B (nominal), Big Group and Bigger Group (ordinal), and Group A with an average height of 57 inches and Group B with an average height of 51 inches (interval). If you know that Group A is on the average 57 inches tall and that Group B is on the average 51 inches tall, then you already know which group is taller and that they are different from one another.

More questions? See questions #51, #53, and #56.

What Is Reliability, and What Is an Example of How It Can be Established?

Reliability is the quality of an assessment tool such that its results are consistent, and it can take many different forms. Along with validity, reliability is one of two of the most important qualities of a test. If a test is not reliable, it cannot consistently assess a trait, characteristic, or level of performance, and its value as an assessment tool is questionable.

Understanding reliability includes knowing about the three components of any test score. These three components can best be understood by looking at the following equation:

Observed score = True score + Error score

The observed score is the score received on a test or an assessment tool, such as an 89 on a weekly science quiz or a 4 out of 5 on a "How happy do you feel today?" assessment. This is the "real" score—the one that appears on the top of the test when it's returned to you or the one that appears on the computer screen after you've taken the SAT.

The true score reflects the actual level of performance, which is not directly observable. By definition, the true score reflects the actual level of performance without consideration of any other influences.

The error score reflects all those events that account for the difference between an observed and a true score. The error score can consist of errors with their source in the individual (being tired or not studying) or the situation (poor lighting or a very warm room).

If a test were perfectly reliable, then there would be no error score, and the observed score would be perfectly equal to the true score. But given that there are so many factors that affect test performance (and all of those fall under the heading of error score), the equation for reliability looks more like this:

$$\text{Reliability} = \frac{\text{True score}}{\text{True score} + \text{Error score}}.$$

As you can see by the above equation, the lower the amount of error in the equation, the higher the resulting reliability. And, if there is no error whatsoever (a very unlikely occurrence), then reliability would equal the true score, and reliability would be perfect or 100%.

In most cases, reliability is computed using the correlation coefficient, the result of correlating two scores from two different test administrations.

Researchers for the University of Arizona were interested in examining the reliability of the Devereux Early Childhood Assessment (DECA) tool. They did so by looking at the relationships between parent and teacher ratings on a variety of different external scales and scores on the DECA. They found reliable coefficients, but they did not find levels of validity that they deemed satisfactory and therefore called for more research.

Here's the complete reference . . .

Otilia, C. B., Levine-Donnerstein, D., Marx, R. W., & Yaden, D. B., Jr. (2013). Reliability and validity of the Devereux Early Childhood Assessment (DECA) as a function of parent and teacher ratings. *Journal of Psychoeducational Assessment, 31*(5), 469–481.

More questions? See questions #54, #55, and #59.

What Are the Different Types of Reliability, and When Are They Used?

There are four types of reliability. Each has a particular purpose, and most use some form of the correlation coefficient to establish a numerical index of reliability.

Test-retest reliability examines the stability or consistency of a test over a period of time. It is ideal for establishing the level of reliability of a test that is given on two different occasions.

Parallel forms reliability examines the stability or consistency of a test when the test comes in two different forms. It is ideal for establishing the level of reliability for two forms of the same instrument.

Internal consistency reliability examines whether a test consistently assesses the same dimension or construct. It is ideal for establishing the level of reliability when one is concerned that a test should be measuring one thing, and one thing only.

Interrater reliability examines whether a rating scale is consistent across users. It is established by having two or more users complete the scale and computing their degree of agreement.

What follows is a summary of those qualities along with an example.

Type of Reliability	Purpose	How Is This Type of Reliability Established?	What's a Good Example?
Test-retest reliability	To compute the reliability of a test administered at two different times	Compute the correlation coefficient between time 1 scores and time 2 scores, or r_{time1} r_{time2}.	An assessment of maturity is administered during the fall and spring of junior year in high school. The scores are then correlated to compute the test-retest reliability of the maturity scale.

Type of Reliability	Purpose	How Is This Type of Reliability Established?	What's a Good Example?
Parallel forms reliability	To compute the reliability of a test when two forms are administered at the same time	Compute the correlation coefficient between scores on form 1 of the test and the scores on form 2 of the test, or $r_{form1}.r_{form2}$.	A test of driver readiness is administered in two different forms at the same time to 100 high school students. The parallel forms reliability coefficient is computed to ensure the test is reliable.
Internal consistency reliability	To establish that a test is unidimensional, assessing one, and only one, dimension	Correlate responses to each of the items on the test with the total test score.	A test is designed to examine the construct of attachment. The internal consistentcy reliability is calculated to ensure it measures attachment and nothing else.
Interrater reliability	To establish the reliability of more than one rater	Examine the percent of agreement between different raters who observe the same phenomenon.	A researcher has developed a scale of aggression and wants to make sure that it is reliable. Two raters rate different sets of children on behaviors identified as aggressive. The interrater reliability is the raters' percentage of agreement, or how often they agree.

More questions? See questions #53, #55, and #59.

How Can the Reliability of a Test Be Increased?

If you remember, the formula for reliability is

$$\text{Reliability} = \frac{\text{True score}}{\text{True score} + \text{Error score}},$$

and the lower the amount of error score in the formula, the closer the reliability coefficient is to the true score and the more perfect the assessment.

You can see in the equation below that error score is composed of two elements: trait error and method error.

$$\text{Error score} = \text{Trait error} + \text{Method error}$$

Trait error is due to individual differences between test takers, such as the amount of time they study, how well prepared they are in general, their state of health, and their motivation. Method error is due to those differences that are not characteristic of an individual, such as the physical properties of the test, the location of the test, and how comfortable test conditions are.

The most useful way to increase reliability is to decrease error score, which means minimizing trait and method error as much as possible. From a pragmatic point of view, it is easier to minimize method variance than to minimize trait variance. For example, it is much easier to ensure that test instructions are clearly written and the test location provides adequate light than it is to decrease a test taker's anxiety.

Following are some possible sources of trait error:

- Ill health
- Poor preparation
- Lack of motivation
- Lack of interest

And here are some possible sources of method error:

- Poor instructions
- Illegible items
- Poorly constructed items
- Unfamiliarity with testing format
- Dependency of items (they can't stand alone)
- Implausible alternative answers

Finally, increasing the length of a test can increase reliability as well, because the larger the number of test items, the more accurately the test samples the universe of all test items. In this way, the observed score can get closer to the true score.

More questions? See questions #53, #54, and #59.

What Is Validity, and What Is an Example of How It Can Be Established?

Validity, after reliability, is the second most critical psychometric property of any test, scale, or instrument that assesses behavior. It is the property of an assessment tool that indicates that the tool does what it says it does. And if the test or assessment tool does have validity, then outcomes associated with it have meaning. Without validity, it is impossible to attach meaning to the outcome score, and the results are in many ways useless. And of course, if the instrument that is used does not have validity, then the value of the experiment of which it is a part is called into question as well.

Basically, when one tries to establish the validity of a test, the presence of external evidence is critical. For example, if a judge deems that the test items sample what is supposed to be tested, that's evidence for a particular type of validity. Or, if a score on an established reliable and valid scale relates significantly to the scores of the scale under development, that is further evidence that the test being developed works as it should. However, because we can't attach a number to the notion of validity very easily, we talk about a degree of validity that rests along a continuum from weak to strong.

For example, for a survey of attitudes toward the use of public funds to support building professional sports facilities, the developer of the survey may use a group of architects, developers, sports fans, and other related audiences to examine the possible survey items and judge whether the items accurately reflect what the survey is aimed at assessing.

As an example, a group of researchers investigated the reliability and the validity of the patient-rated Migraine Treatment Optimization Questionnaire (M-TOQ). The researchers had almost 300 patients complete a series of other scales and tests that were thought to be related to the presence of migraine headaches in addition to completing the test under development. Study participants completed questionnaires and scales, previously validated and in use, such as the Headache Impact Test, which measures the impact of migraines on quality of life across multiple dimensions of everyday living. The goal was to see how established indicators of migraine headaches reflected the reliability and validity of the scale under development. This reflects one of the most common practices for establishing the validity of a new assessment tool—see how the results of the new test relate to those of an established, and valid, test.

Here's the complete reference . . .

Lipton, R. B., Kolodner, K., Bigal, M. E.,Valade, D., Láinez, M. J. A., Pascual, J. . . . Parsons, B. (2009). Validity and reliability of the Migraine-Treatment Optimization Questionnaire, *Cephelalgia, 29*(7), 751–759.

More questions? See questions #51, #57, and #58.

What Are the Different Types of Validity, and How Are They Computed?

There are three types of validity, and all are used to assess the "truthfulness" of different types of assessment instruments to determine whether they do what they are designed to do.

Content validity is used to examine whether an assessment tool covers the universe of items it purports to measure. Content validity might be used to evaluate an achievement test.

Criterion validity is used to examine whether an assessment tool relates knowledge, skills, or abilities to other criteria that are related to the test under development. There are two types of criterion validity. Concurrent validity is used to assess the current status of an assessment tool (such as how well a newly developed test of spatial skills correlates with the ability to solve block puzzles), and predictive validity is used to assess the future value of an assessment tool (such as how well a newly developed test of interpersonal relationships predicts future physicians' bedside manner).

Construct validity is used to examine whether an assessment tool reflects some underlying psychological construct such as happiness, aggression, hope, or optimism. It is established by examining the relationship between scores on the newly developed test and performance on other tasks that theoretically reflect the construct under study.

Here's a summary of these different types of validity, when they are used, and how they are established.

Type of Validity	When You Use It	How You Do It
Content validity	When you want to know whether a sample of items reflects an entire universe of items	Ask an expert to judge whether the test items reflect the entire universe of items being measured.

Type of Validity	When You Use It	How You Do It
Criterion validity	When you want to know whether test scores are systematically related to other criteria	Correlate the scores from the test with some other measure that is already validated and that assesses the same set of abilities.
Construct validity	When you want to know whether a test measures some underlying psychological construct	Correlate the set of test scores with some theorized outcome that reflects the construct for which the test is being designed.

More questions? See questions #56, #58, and #59.

How Can the Validity
of a Test Be Increased?

Increasing the reliability of a test is fairly straightforward—reduce the trait or method error or increase the number of items on the assessment tool. Increasing the validity of a test is not quite as straightforward but, interestingly enough, relates closely to changing the reliability of an assessment tool.

To increase the validity of a test, you can consider the following strategies.

First, while increasing reliability will not necessarily increase validity, reliability is an important precondition for validity, so make sure that the test is reliable. A low reliability coefficient constrains what the highest validity coefficient can be, so the higher the reliability, the more room for validity to increase.

Second, confirm that what you are testing is what you want to test. In the area of achievement tests, some people use a table of specifications that consists of two axes, one that defines skills and one that defines content. The task is to make sure that the newly created items accurately reflect the intersection of a skill (such as memorization) with content (such as the periodic table).

Third, if the test does not seem to be functioning as desired, try reworking items to be sure that they tap the content, construct, or skill that the test is designed to assess.

Fourth, as with reliability, developing the test based on a larger sample of test takers helps develop more focused and more accurate items that better reflect the content or constructs being assessed. The more varied the sample, the more likely it is that the items developed, after revision, will reflect the skills and abilities and knowledge that are being assessed.

Finally, be sure to pretest any tests, scales, assessment tools, and other types of outcome measures to confirm that all the qualities of a "good" test are there. These include clarity of instructions, ease of use, approachability by a variety of test takers, and understandability. All of these factors can contribute to lower reliability and constrained validity.

More questions? See questions #56, #57, and #59.

What Is the Relationship Between Reliability and Validity?

As you know by now, reliability is the quality of a test that speaks to its consistency and stability. If a test is reliable, several different administrations will result in a similar level of performance relative to the entire group's performance. In addition, you know that validity is the quality of a test that speaks to its veracity, truthfulness, or authenticity. If a test is valid, it assesses what the test was designed to assess.

But just as important as these two ideas to ensuring that an assessment tool works as it should is the relationship between reliability and validity.

It is entirely possible to have a test that is reliable but not valid. However, you cannot have a valid test without it first being reliable. For example, a test can do whatever it does over and over (demonstrating reliability) but still not do what it is supposed to (oops—no validity). But if a test does what it is supposed to (it is valid), then it has to do it consistently (and be reliable).

For example, here is a multiple-choice item from a test.

1. During the summer of 1776, where did the Continental Congress meet?

   a. Washington, D.C.

   b. Philadelphia

   c. New York

   d. Boston

If one were to compile 50 items like this and administer this test at two points in time, the test-retest reliability would probably be adequate. But, if one were to call this a test of international studies or introductory psychology—which it is surely not—it has no validity for those purposes. So, even though the results are replicable (hence, reliable), the test is not valid. On the other hand, if this were examined as a test of introductory

American history and the content validity were established, then there would be a clear path to establishing its reliability as well.

One more important point is that both reliability and validity are essential qualities of testing tools so that the results of hypothesis testing that uses such tools have credibility. If a test is not reliable (and hence not valid) or simply not valid, any one test of a hypothesis will provide results that do not fairly reflect the integrity of that hypothesis and the research question from which it came.

Finally, and on a bit more technical note, the maximum level of validity is equal to the square root of the reliability coefficient. For example, if the reliability coefficient for a test of mechanical aptitude is .87, the validity coefficient can be no larger than .93 (which is the square root of .87). What this means is that, again, the validity of a test is constrained by how reliable it is. And that makes perfect sense if we stop to think that a test must do what it does consistently before we can be sure it does what it says it does.

More questions? See questions #51, #53, and #56.

PART 7

UNDERSTANDING THE ROLE OF HYPOTHESES IN STATISTICS

What Is a Hypothesis, and Why Is It Important in Scientific Research?

A hypothesis (plural *hypotheses*) is an educated guess. It is only one part, albeit a very important one, of the scientific process. There are several kinds of hypotheses, which we will deal with in later questions in this section of *100 Questions (and Answers) About Statistics*.

The hypothesis's most important role is to take a research question such as "Do boys and girls differ in their level of math and verbal achievement when they enter high school?" and allow it to take the form of a declarative statement that is testable, such as "Boys and girls differ in their math and verbal achievement when they enter high school."

This may seem like a trivial difference, but it is not. Hypotheses allow questions to be stated as actions—ideas to be acted upon.

The best way to understand the importance of a hypothesis is to list the steps of the scientific process and briefly discuss how each bears on the use of this tool.

1. Ask the research question.

2. Identify the important factors in the research question.

3. Formulate and state a hypothesis.

4. Collect data that are relevant to the hypothesis.

5. Test the hypothesis.

6. Revise/Consider the hypothesis.

7. Revise/Consider the theory.

8. Ask new questions.

While each of these nine steps does not directly deal with the formulation and testing of a hypothesis, the nature of the hypothesis is affected by each one.

Steps 3, 5, and 6 above are especially important to understand.

Step 3, formulate and state a hypothesis, allows the researcher to phrase in testable terms the original question that was asked as part of the research project. The curiosity that scientists have about their work leads to excitement and great expectations about the importance of the resulting findings. But, these findings have to be part of a systematic process in which questions are framed in such a way that they can be answered, with the answer to each question contributing something (small or large) to the answer to the overall question being asked.

Step 5, test the hypothesis, is where those questions become articulated as statements about what variables are involved and what the researcher expects to see as an outcome. For example, if math and verbal skills are being examined in a group of children who are placed in groups as a function of gender, the variables of gender, math achievement, and verbal achievement are specified—and their relationship further so—by the nature of the hypothesis.

Step 6, revise/consider the hypothesis, is where the researcher considers and reconsiders the theory on which the hypothesis is based and the hypothesis itself, given the results of the test of the hypothesis. In this circular process, feedback results in a new, more precise test of the original hypothesis as an effort continues to answer the original questions asked at the beginning of the process.

More questions? See questions #61, #64, and #65.

What Are the Characteristics of a Good Hypothesis?

A well-written and well-thought-out hypothesis can make all the difference between a successful and unsuccessful research effort. This is primarily because a well-written hypothesis reflects a well-conceived research project based on an adequate review of the literature and a logical proposition about the relationship between variables.

Here is a summary of the characteristics of a good hypothesis.

First, a good hypothesis is stated in declarative form and not as a question. For example, "Are retention rates for first-year students at state universities low because students run out of money?" could, with some review of the literature, become, "Retention rates for first-year students at state universities are lower than the average because students cannot afford to return for the second semester due to a shortage of funds." The hypothesis becomes a direct and clear statement.

Second, a good hypothesis proposes a relationship between variables. In the example we just provided, the variables are whether or not the student remains in school (retention) and the reason for not remaining in school if the student leaves. In this example, the idea that is being tested is that new students do not remain in school because school becomes too expensive.

Third, a good hypothesis reflects the literature or the results of previous studies on which the hypothesis is based. This is where good old-fashioned detective work at the library or online provides the information needed to best understand the possible relationships that might be found and their importance to the overall research mission.

Fourth, a good hypothesis is brief and to the point. It is not a review of the literature or a rationale for the hypothesis itself. Rather it is a concise and clear statement of the relationship between variables such that any other person with some familiarity with the subject matter could read the hypothesis and fully understand the central purpose of the research study.

Finally, a good hypothesis is testable. The variables are clearly understood, as is their proposed relationship. In our example, the central question is the relationship between continued enrollment in school and why that may not occur. The hypothesis narrows that question to look specifically at one reason why continued enrollment may not occur. Given the way the hypothesis is stated, it allows the question to be tested and the results and new knowledge gained to be applied to the next hypothesis and subsequent testing.

More questions? See questions #60, #63, and #64.

How Do a Sample and a Population Differ From One Another?

One of the major functions that inferential statistics plays is to allow testing of a hypothesis with a sample and then using the results from that sample to infer how well they apply to a population. A sample is simply a subset of a population. The idea of using a sample rather than a population has some very sound reasoning behind it.

Since a sample is smaller, assessing some outcome requires fewer resources such as money, personnel, facilities, and so forth. For example, if we are interested in the height of sixth graders in a large, urban school district with 10,000 sixth graders, we could (if we do so correctly) measure only 100 or 200 students and get a fairly accurate measure of the average height of all sixth graders.

How do we know that our assessment based on a much smaller number of participants is accurate and accurately reflects the population values?

A measure of how well a sample approximates the characteristics of a population is called sampling error. Sampling error is basically the difference between the values of the sample (called a sample statistic) and the values of the population (called a parameter parameter). The higher the sampling error, the less precision you have in sampling and the more difficult it will be to make the case that what was found in the sample indeed reflects what you expect to find in the population. The job of any researcher is to minimize sampling error in an effort to get the most accurate representation of the population values as possible.

More questions? See questions #60, #61, and #65.

What Is a Null Hypothesis, and How Is It Used?

The null hypothesis (*null* meaning "none" or "void") is a statement of equality. It is the conceptual starting point that reflects no relationship exists between variables. Why? Because given no other knowledge of the relationship between variables, the null hypothesis is the only logical place to begin a research effort.

A null hypothesis can take this form:

$$H_0: \mu_1 = \mu_2,$$

where

H_0 equals the research hypothesis,

μ_1 equals the value of the population parameter for group 1, and

μ_2 equals the population parameter for group 2.

Note that the above example, that the null hypothesis is that two group averages are equal, is only one of many examples of the null.

For example, the following null hypothesis assumes no relationship between participation in a summer internship program and job satisfaction: "There will be no difference in the satisfaction level of those first-year employees who have had summer internships and those who have not." As you notice, no relationship is stated between participation and job satisfaction—the beginning assumption here is that both those who had an internship and those who did not will have equivalent or equal scores on a measure of job satisfaction.

The null hypothesis is an important tool for two reasons.

First, it is a starting point because it is the state of affairs in a relationship that is most plausible given no other information. The most we can say about the relationship between any two or more variables, given no other information, is that they are unrelated to one another. By starting our serious investigation of any hypothesis assuming that all things are equal, we start at an unbiased and clearly defined place.

Second, the null hypothesis is a benchmark that we will eventually use for comparison purposes when we do collect information that is relevant to the question being asked. Once we know the starting point (the null), we then have something to which we can compare other findings and outcomes. In this way, we can see whether the null is the most reasonable explanation for any differences we may see in the test of the relevant hypothesis.

Finally, it is important to note that the null hypothesis is as much a conceptual starting point as anything else. It rarely finds its way into journal articles and research reports. However, it is a direct reflection of the research hypothesis, and the null is always implied whenever any research question is asked.

More questions? See questions #60, #62, and #64.

What Is a Research Hypothesis, and How Is It Used?

The null is a statement of equality—a starting point at which there is no relationship between variables. The research hypothesis, in contrast, is a statement that there is a relationship between variables. That relationship can take many different forms. Most importantly, research hypotheses are statements of inequality.

For example, a research hypothesis might posit that a difference in income exists between homeowners and renters, a difference exists between urban and rural residents in their attitude toward recycling, or a correlation (or relationship) exists between years playing contact sports and head injuries. In all these cases, given the information that we have (such as a review of the literature or the results of previous research), a difference is hypothesized.

There are two types of research hypothesis: nondirectional and directional.

A nondirectional research hypothesis reflects a difference between groups or a relationship between variables, but it does not indicate the direction of that difference. For example, the research hypothesis that "There is a difference between homeowners and renters in their monthly income" tells you nothing about whether homeowners or renters have more or less income than the other group—just that there is a difference.

A research hypothesis such as this might appear in a research article or a report as

$$H_1: \bar{X}_{owners} \neq \bar{X}_{renters,}$$

where

H$_1$ represents the research hypothesis (there can be more than one research hypothesis),

$\neq$ means "is not equal to,"

$\bar{X}_{\text{owners}}$ represents the average income of homeowners, and

$\bar{X}_{\text{renters}}$ represents the average income of renters.

A directional research hypothesis reflects a difference between groups or a relationship between variables, and it indicates a direction for that difference. For example, the research hypothesis that "The monthly income of homeowners is higher than the monthly income of renters" tells you quite specifically what the researcher expects as an outcome.

A research hypothesis such as this might appear in a research article or a report as

$$H_1: \bar{X}_{\text{owners}} > \bar{X}_{\text{renters,}}$$

where

H_1 represents the research hypothesis,

> means "greater than,"

$\bar{X}_{\text{owners}}$ represents the average income of homeowners, and

$\bar{X}_{\text{renters}}$ represents the average income of renters.

The > or "greater than" sign in this research hypothesis is only one of many operators that can indicate a relationship between variables. For example, a nondirectional hypothesis could be stated as

$$H_1: \bar{X}_{\text{owners}} \geq \bar{X}_{\text{renters,}}$$

where homeowners' income is expected to be greater than or equal to renters' income.

More questions? See questions #60, #61, and #63.

How Do the Null and Research Hypotheses Differ From One Another?

The null and research hypotheses differ from one another in several very important ways.

First, the null hypothesis is a statement of equality, while the research hypothesis is a statement of inequality.

Second, null hypotheses are typically unstated in journal articles and research reports, while research hypotheses are explicitly stated early in an article or report.

Third, null hypotheses refer to populations, while research hypotheses refer to samples. A sample of participants from the population is selected, and the research hypothesis is tested using that sample. Then, the results can be inferred or generalized to the population.

Fourth, null hypotheses are always stated using population parameters, as in this example:

$$H_0: \mu_1 = \mu_2$$

This means that the average of population 1 equals the average of population 2. The μ character represents the population parameter for the average.

On the other hand, research hypotheses are always stated using sample statistics. Here's an example:

$$H_1: \bar{X}_1 \neq \bar{X}_2$$

This means that the average of sample 1 does not equal the average of sample 2.

Fifth, because the entire population cannot be tested (it is impractical both financially and strategically), it is impossible to say with 100% certainty whether the null hypothesis is true or false. For example, even though we may have observed a difference between one sample and another sample in average scores on a test, it is only with a certain degree of confidence (albeit very high in many cases) that we can say that the results from the sample apply to the population. It is for this reason that we say the null hypothesis is indirectly tested and the research hypothesis directly tested.

More questions? See questions #60, #61, and #62.

PART 8

UNDERSTANDING THE NORMAL CURVE AND PROBABILITY

Why Is Probability Important to the Study of Statistics?

Much of what you learn in studying statistics has to do with the normal or the bell-shaped curve.

Understanding the normal or bell-shaped curve allows us to understand how a probability, or the likelihood of an occurrence, can be associated with any outcome. For example, how likely is it that a student will receive a score of 87 on a test on which the class average was 93? Or, how likely is it that sales in the Midwest office of a national real estate company are representative of sales at all offices throughout the nation?

Assigning a probability to an outcome allows us to come up with answers to such questions.

We can determine whether the probability of an outcome is high or low, and then we can make a decision, given certain rules, about whether we find that probability level acceptable.

The study and use of probability also allow us to determine the degree of confidence that we have in stating that a particular outcome is "true." For example, if we observe that men differ from women in their level of aggression, how confident can we be that such a finding is "true"? Perhaps it is just the result of a poorly designed experiment, or maybe it is a chance occurrence due to such factors as sampling error, because the sample does not do a very good job of representing the population.

Finally, the whole notion of probability is closely tied to the role of null and research hypotheses. The research hypothesis is tested using a sample of participants from a much larger population. Understanding the role of probability allows us to apply those findings to the population at large, but with one important caveat. Since we cannot test the population directly, we are taking a very well-educated guess as to how closely the sample's findings can be applied to the population and with what degree of certainty.

More questions? See questions #67, #69, and #72.

What Is the Normal or Bell-Shaped Curve?

The normal or bell-shaped curve, as you see in Figure 67.1, is a visual representation of a distribution of data that has three very special characteristics. Most important, because of these characteristics, the normal curve becomes the basis for much of how inferential statistics operates.

In this example, the *x*-axis represents values (such as how smart people are), and the *y*-axis represents the frequency or probability of those values (lots of people or very few people and everything in between).

Figure 67.1 The normal curve

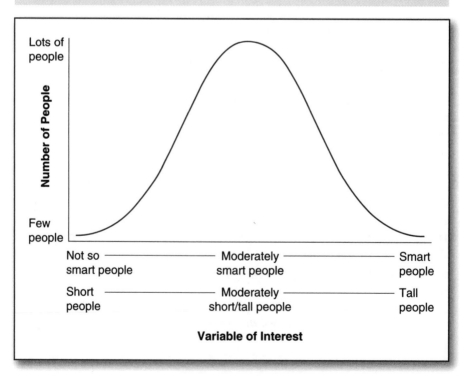

The first key characteristic is that the mean, median, and mode are all equal to one another. The central-most point in the distribution of scores represents the mean, the median, and the mode.

Second, the entire set of scores is symmetrical around this one central point. One side of the curve is the mirror image of the other, and the two halves are identical to one another.

Finally—and this is the characteristic that has the most important implications for much of what we will be discussing in this section of *100 Questions (and Answers) About Statistics*—the tails of the curve are asymptotic. This means that even though each tail of the curve gets closer and closer to the *x*-axis, the tail never will actually touch or meet the *x*-axis.

The importance of this characteristic is that there is always a likelihood represented by some value on the *y*-axis (even if that likelihood is very small) that any *x*-value can occur. In other words, regardless of how extreme an *x*-value is, it may be found in one tail of the curve or the other. The likelihood of a given *x*-value's occurrence becomes very important in our later discussions.

More questions? See questions #66, #68, and #73.

What Are Skewness and Kurtosis, and How Do Distributions Differ on These Characteristics?

Not all distributions of scores are perfectly normal. In fact, while the distribution of many scores approaches a normal shape, many sets of scores are distributed in other ways.

Sometimes a distribution is skewed, showing a lack of symmetry such that one tail of the distribution is longer than the other tail. You can see an example of this in Figure 68.1.

Figure 68.1 How a distribution of scores can be skewed

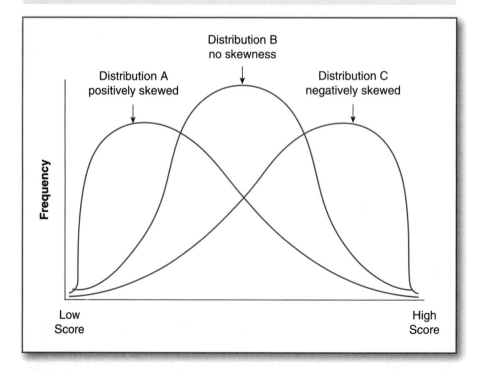

If the left-hand tail of a distribution is longer than the right, as when there are many higher values than lower values, then the distribution is negatively skewed. An example might be the distribution of measurements of height for 100 adults, including 75 players from the National Basketball Association, all of whom are exceptionally tall.

If the right-hand tail of a distribution is longer than the left, as when there are many more lower values than higher values, then the distribution is positively skewed. An example might be the distribution of scores on a test for which only one quarter of the class was given the proper materials to prepare, meaning that most of the test takers performed more poorly than one would expect.

The second way a distribution of scores can differ from others is in its degree of flatness or kurtosis, as can be seen in Figure 68.2.

Figure 68.2 How distributions of scores differ in kurtosis

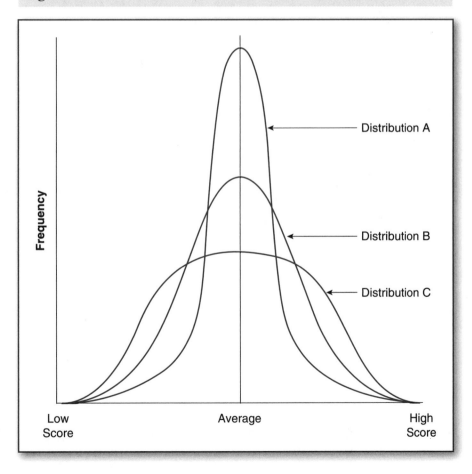

As you can see, some distributions are "flatter" than others. A distribution that is leptokurtic (distribution A in Figure 68.2) is one in which scores tend to group around the middle of the distribution and fewer scores are in the extremes. This might be the case when the variability of a set of scores is relatively low, such as when everyone scores very close to the average on an exam.

In a platykurtic distribution, or distribution B in Figure 68.2, the distribution of scores is relatively flat compared to a normal distribution, and the amount of variability tends to be spread equally across all possible outcomes. For example, the distribution of scores on a final exam on which everyone does about equally as well (or equally as poorly) would be flatter than a distribution of scores that are normally distributed or of scores that are mostly very close to the mean.

More questions? See questions #66, #67, and #73.

What Is the Central Limit Theorem, and Why Is It Important?

In question #67, you saw an illustration of what the normal curve looks like, and you will shortly learn how many of the basic concepts underlying descriptive and inferential statistics are based on the shape and the properties of this curve.

But what if the distribution of scores is not normal—whether it be skewed or some other shape? Do these same rules of inference apply? Yes, they do. Because of the central limit theorem, one can apply the rules of inference to almost any distribution of scores.

The central limit theorem proposes that even in a set of scores that is distributed in a non-normal fashion, repeated sampling from the distribution will produce an average of the samples that is distributed normally.

For example, let's assume we have a set of 100 scores (ranging from 1 to 5) with associated frequencies (for example, a score of 1 occurs 25 times), and we plot these scores as you see in Figure 69.1.

Figure 69.1 A non-normal distribution of 100 scores

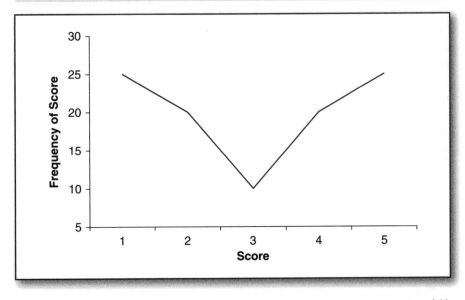

Now, let's take a random sample of size 5 from this set of 100 scores and compute the mean. Then, let's take another random sample . . . until we have taken hundreds of samples of size 5, computed the mean for each one, and then plotted those values as you see in Figure 69.2.

Figure 69.2   A normal distribution of means

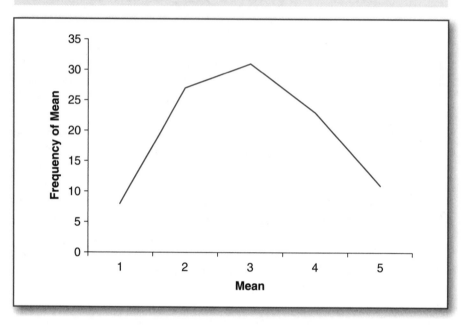

As you can see, this new distribution assumes many of the familiar qualities of the normal curve, including a somewhat bell shape, symmetry, and tails that would not touch the x-axis even if the number of means selected were infinite.

The lesson here is that regardless of the shape of the parent distribution (most often the population from which the sample is selected), the shape of the sample means will approach normality, and this is what's needed to make any sense out of the inferential method.

More questions? See questions #67, #72, and #73.

What Is a *z* Score, and How Is It Computed?

There are so many different studies that if we want to compare performances on a certain variable from different distributions, we need a metric with some common basis. Standard scores allow us to do just that. These scores are comparable to one another because they all are expressed in units of standard deviations.

In the social and behavioral sciences, as in other fields, the most frequently used standard score is the *z* score. The *z* score is called a variety of other names, such as *z* value, normal score, and standardized variable. And there are other types of standard scores, such as *T* scores.

Here is the formula for computing a standard score:

$$z = \frac{X - \bar{X}}{s},$$

where

z is the *z* or standard score,

X is the value for which you want to compute the *z* score,

$\bar{X}$ is the mean of the sample, and

s is the standard deviation.

For example, if the average score on a test is 78 and the standard deviation is 3, then a *z* score for a raw score of 81 is

$$z = \frac{81 - 78}{3} = +1.0$$

If the raw score were 77, then the *z* score would be

$$z = \frac{77 - 78}{3} = -0.33$$

Here's a set of raw scores with their corresponding z scores. The mean is 5.6, and the standard deviation is 2.1.

Raw Score	z Score
5	−0.29
7	0.68
6	0.19
4	−0.77
5	−0.29
6	0.19
1	−2.23
8	1.16
6	0.19
8	1.16

We'll shortly get into why z scores are important in the world of descriptive and inferential statistics, but for now, note the following:

- Those scores that are above the means, such as 6 and 7, have a corresponding positive z score. Those scores below the means, such as 1 and 4, have a corresponding negative z score.
- Positive scores always fall to the right of the mean (or in the right-hand side of the distribution), while negative scores always fall to the left of the mean (or in the left-hand side of the distribution).
- Stating that a raw score has a corresponding z score of 1 is the same as saying that the raw score is 1 standard deviation above the mean.
- z scores across different distributions are comparable. So, a z score of 1.237 from a distribution with a mean of 100 and a standard deviation of 22 is the same as a z score of 1.237 from a distribution with a mean of 55.4 and a standard deviation of 4.3. Both scores are 1.237 standard units, standard deviations, or z scores from the mean.

More questions? See questions #71, #72, and #73.

How Can Excel Be Used to Compute z Scores?

Using Excel to compute z scores is a simple procedure that can be done by following these steps. After completing the four steps below, your worksheet should look as it appears in Figure 71.1.

Figure 71.1 Using Excel to compute z scores

	Raw Score (X)	z Score
	89	0.68
	78	−0.16
	49	−2.37
	85	0.37
	93	0.98
	68	−0.92
	79	−0.08
	90	0.75
	82	0.14
	88	0.60
Average	80.1	
Standard Deviation	13.1	

1. List all the raw scores in a column in a new worksheet. Label the column "Raw Score (X)," as you see in Figure 71.1.

2. Using the =AVERAGE function, compute the mean and place it at the bottom of the column.

3. Using the =STDEV.S function, compute the standard deviation and place it below the mean. You can see steps 2 and 3 completed in Figure 71.1.

4. In cell C2, place the following formula . . .

$$z = (B2 - \$B\$12)/\$B\$13$$

and then copy that formula down from cell C2 through cell C11.

This formula placed in cell C2 subtracts the raw score (which in this case is 89) from the mean (80.1) and then divides that product by the standard deviation (13.1). The dollar signs are present so that when the formula is copied down the column, the values for the mean and the standard deviation are constant and not relative.

More questions? See questions #70, #72, and #73.

What Is the Relationship Between *z* Scores and the Normal Curve?

The normal curve has the three characteristics that we already mentioned: the mean, median, and the mode being equal to one another; both sides being symmetrical; and the tails never touching the *x*-axis.

But it has another set of characteristics that are very important to our discussion of *z* scores and the idea of inference and the use of inferential statistics.

The normal curve can be divided into segments, as you see in Figure 72.1, and you can see that a specific percentage of outcomes falls within each of these areas.

Figure 72.1 The normal curve and associated percentages

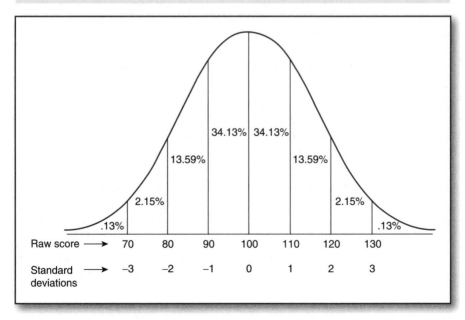

		34.13%	34.13%			
13.59%					13.59%	
2.15%						2.15%
.13%						.13%

Raw score ⟶ 70 80 90 100 110 120 130

Standard ⟶ −3 −2 −1 0 1 2 3
deviations

For example, between the mean and 1 z score (or standard deviations), you will find 34.14% of all scores. If the average score on some assessment were 100 and the standard deviation were 10, then 34.14% of all the scores in the distribution would fall between a score of 100 and 110. Here's a summary table that shows you exactly what percentage of scores fall between what points under the normal curve as a function of their distance from the mean.

The distance between . . .	Includes this percent of all scores . . .
the mean and 1 standard deviation	34.13%
1 and 2 standard deviations	13.59%
2 and 3 standard deviations	2.15%
3 standard deviations and above	0.13%

This is the case for negative z scores as well. For example, 34.13% of the cases under the normal curve always fall between the mean and $-1z$ or 1 standard deviation below the mean. This is because the normal curve's two halves are symmetrical and any one half accounts for exactly 50% of the scores under the curve.

Why is this so important? Because it is a very small step from what you just read and the assignment of probabilities associated with any outcome within a distribution of scores.

In other words, for a set of scores with a mean of 100 and a standard deviation of 10, the probability that a raw score between 100 ($z = 0$) and 110 ($z = +1.00$) will occur is 34.13%—it falls within the area under the curve encompassing 34.14% of all scores. Or, for a set of scores with a mean of 100 and a standard deviation of 10, the probability that a raw score between 80 ($z = -2.00$) and 90 ($z = -1.00$) will occur is 13.59%.

Though it is beyond the scope of what we are doing here, be aware that specific tables can be used to assign the exact probability of any outcome or the likelihood that a specific score will occur. For example, given a z score of 1.65, the probability that a score will fall below that in a normal distribution is about 95%, and the probability that a score will fall above it is 5%.

This notion of assigning probabilities to outcomes becomes very important as the basis for many of the inferential tests we will shortly discuss.

More questions? See questions #66, #67, and #69.

How Do *z* Scores Relate to Hypothesis Testing?

The most important reason why inferential statistics are such powerful tools is that they allow us to use a representative sample taken from a population and infer, from the results of a test or measure of that sample, something about the population. That's big.

But also very important is the fact that we can assign probabilities to any outcome (such as, but not limited to, *z* scores) taken from a sample. Then, using a set of rules, we can decide whether that outcome is likely enough for us to attribute the finding to some known event (such as the use of a treatment to help children read better) or we must simply attribute it to chance or error. *z* scores or standard scores, our first example, are representative of many different types of scores—all of which have associated outcomes or probabilities.

For example, if you were interested in whether a coin you had was a true and fair coin and not rigged to come up with an unusual number of heads, you might make the following argument: If you tossed that coin 10 times, then by chance alone, you could expect there to be 5 heads and 5 tails on those 10 flips. The probability associated with any one flip is .5 (either a head or a tail).

But, on 10 consecutive flips, how confident could you be that you would see 7, 8, or 9 heads (or tails?). Well, the probability of getting 8 heads on 10 flips is .04 or 4%, and the probability of getting 9 heads on 10 flips is .01 or 1%, and the probability of getting all heads on 10 successive flips is really small (less than .001 or .1%).

So, just how extreme does the result of tossing a coin 10 times have to be for you to say that the coin is not fair or true? That's your decision. In most cases, a criterion of .05 or 5% would work just fine. In other words, if the probability of an outcome is less than 5%, one might say that the outcome is so unlikely, it cannot be due to chance but must be due to something other than chance—in this case, a fake or trick coin. So, if you get 10 successive heads on 10 successive flips, there is a very high likelihood that you have a fake coin.

z scores are our first introduction to how assigning probabilities to outcomes can help us make decisions about whether those outcomes are due to something systematic or just due to chance. We'll explore a host of those possibilities throughout the next set of questions.

More questions? See questions #66, #69, and #72.

PART 9

UNDERSTANDING THE CONCEPT OF SIGNIFICANCE

How Does Inference Work, and What Is an Example?

Inference is the process through which scientists infer, based on the results from a sample, the characteristics of a population. Its success depends upon how accurately the sample represents the population, how reliable and valid the instrumentation is, and other factors. When done correctly, the inferential method is a very powerful and efficient tool.

First, let's look at an example of a research project to examine the steps used in the inferential method.

Luis Facorro and Melvin Defleur examined differences between American and Spanish persons in their recall of different news stories. Participants were exposed to one of three local news stories presented via newspaper, computer screen, television, or radio. In sum, the researchers found differences in recall levels between cultures as well as across the media in which the news stories were presented.

Basically, there are four steps to the inferential method, and these are exemplified by the above study:

1. The researchers selected representative samples of 720 communication students who would participate in the study. The idea was that these 720 persons constituted a sample that represented the population of both Spanish and American students studying communications.

2. Each participant was exposed to various news stories under one of four conditions (newspaper, computer screen, television, or radio). Then recall tests were administered to measure how accurately the participants could recall story details under each condition. The mean scores for the groups were computed and compared.

3. A conclusion was reached as to whether the difference between the scores was the result of chance (meaning some factor other than cultural or presentation differences) or the result of "true" and statistically significant differences between groups (meaning the results were due to conditions and/or culture).

4. A conclusion was reached as to the relationship between culture and condition and recall. In other words, an inference, based on the results of an analysis of the sample data, was made about the population of all communication students.

Here's the complete reference . . .

Facorro, L. B., & Defleur, M. L. (1993). A cross-cultural experiment on how well audiences remember news stories from newspaper, computer, television, and radio sources. *Journalism & Mass Communication Quarterly, 70*(3), 585–601.

More questions? See questions #75, #77, and #78.

What Is the Concept of Significance, and Why Is It Important?

Understanding the concept of significance can be a bit challenging, but it is basically as follows: Statistical findings are significant if the outcomes of an experiment are more likely to occur because of what the experimenter did than due to chance.

You may remember that in question #63, we made the point that the null hypothesis is important because it is a starting point where we assume that there is no relationship between variables, no differences between groups, etc. In other words, unless we can explain why there may be an observed difference between groups, we assume that any difference we find is due to chance factors.

For example, let's look at two groups of middle-aged males and an experiment that involves a weight reduction program. Group 1 receives the treatment, which is a special form of exercise. Meanwhile, group 2 receives no treatment at all; they are the control group. At both the beginning and the end of the experiment, weights are compared. The research hypothesis is that men who participate in the special exercise or treatment group will lose more weight than the group of men who do not participate in any special exercise program.

The results are in, and there is a difference between the two groups. Let's say that the exercise group lost 5 pounds and the nonexercise group lost 1 pound. The big question is this: Is the difference between the two groups unlikely enough to have occurred so that we can attribute the difference to the weight loss program, or is it just due to chance such as sampling error or poor selection of participants or a host of other factors?

Here's where we apply the rules of inference. If the difference between the two groups is 3 pounds or greater, we will say with some degree of confidence (let's say 95% confident) that the reason for the difference is the treatment and nothing else. If the difference is less than 3 pounds, we will say the difference is not large enough for us to say that it is statistically significant. In this case, the weight loss is 4 pounds, exceeding what we would expect by chance, and the findings are significant.

Technically, significance is the probability that on any one test of the null hypothesis (that there is no difference between the weights of the two groups), we will falsely conclude that there is a difference, when in fact there is none.

Traditionally, the level of significance is set at .01 and .05, or 1% to 5%, and it is here where we assign a probability to an outcome (called the obtained value) based on a certain statistical test. Then we decide whether the findings are "significant" (or not due to chance) or due to chance.

More questions? See questions #74, #76, and #77.

What Are a Type I and a Type II Error?

A Type I error (also called significance level or alpha, represented by the Greek letter α) is the probability of rejecting a null hypothesis when it is indeed true. A goal of our work is to minimize that error. Remember that since the null hypothesis is based on the population, which is never directly tested, we never really know whether the null is true or false.

For example, let's say that we are reasonably confident that there is a difference between males and females in their verbal SAT scores. The null state reflected by the hypothesis is that there is no difference between the two groups. We test the research hypothesis that there is a difference and find out that, at the .01 level of significance, the observed difference is significant. We are confident at the .01 percent level that the difference occurred as a function of gender. The level of a Type I error is under the control of the researcher and is set by the researcher, usually at .01 or .05. That's the level at which the research hypothesis will be tested.

Significance level is often expressed as, for example, $p = .37$ (where the exact level of a type I error is .37), $p < .05$ (where the unspecified level of a Type I error is less than .05), and $p = $ ns (where the level is beyond .05 or whatever level the hypothesis is being tested at with the *ns* sanding for nonsignificant).

A Type II error (also known as beta or β) is the probability that a researcher will accept a null hypothesis when it is actually false. For example, let's say that in reality, the null that there is no difference between males and females is true—there really isn't a difference. A Type II error reflects the probability that we would accept that null hypothesis when indeed it is really false.

The following table summarizes these two types of errors and how they can be controlled.

		Action You Take	
		Accept the Null Hypothesis	**Reject the Null Hypothesis**
True nature of the null hypothesis	**The null hypothesis is really true.**	1 ☺ Bingo! You accepted a null when it is true and there is really no difference between groups.	2 ☹ Oops—you made a Type I error and rejected a null when there really is no difference between groups. Type I errors are also represented by the Greek letter alpha, or α.
	The null hypothesis is really false.	3 ☹ Uh-oh—you made a Type II error and accepted a false null hypothesis. Type II errors are also represented by the Greek letter beta, or β.	4 ☺ Good job! You rejected the null hypothesis when there really are differences between the two groups.

More questions? See questions #74, #75, and #77.

What Are the Steps in Applying a Statistical Test to a Research Hypothesis?

Each research hypothesis has associated with it a particular type of statistic. For example, to test the difference between the average scores for males and females on an exam, one would use a *t*-test for independent means.

And each of these test statistics has associated with it a special distribution, which you compare with the data you obtain from a sample. Comparing the characteristics of the sample to the characteristics of the test distribution allows you to conclude whether the sample characteristics are different from what you would expect by chance.

Here are the general steps to take in the application of a statistical test to any research hypothesis:

1. Make a statement of the research hypothesis.

2. Set the level of significance or Type I error associated with the research hypothesis. This is the degree of risk you are willing to take that you are wrong. The smaller this degree of risk is (such as .01 compared with .05), the less risk you are willing to take. No test of a hypothesis is completely risk-free because you never really know the "true" relationship between variables.

3. Compute the test statistic value or obtained value that is the result of the specific statistical test.

4. Determine the critical value you would expect the test statistic to yield if the null hypothesis is indeed true.

5. Compare the obtained value to the critical value. This is the crucial step. Here, the value obtained from the test statistic (the value you computed) is compared with the value (the critical value that appears in a series of tables) you would expect to find by chance alone.

6. If the obtained value is more extreme than the critical value, the null hypothesis cannot be accepted. Only if the obtained value is more extreme than might be obtained by chance can you reject the null and attribute the findings to the research hypothesis. If the obtained value does not exceed the critical value, the null hypothesis is the most attractive explanation.

More questions? See questions #74, #75, and #76.

What Is the Difference Between Statistical Significance and Meaningfulness?

In general, statistical significance is expressed as a degree of risk (such as $p < .05$ or "p is less than .05"), and it has been the holy grail of research endeavors. But, what tends to be missing from some discussions about statistical significance is the important consideration of the meaningfulness of the data.

For example, the correlation between two variables such as incidence of crime and consumption of ice cream tends to be positive. The more ice cream that is consumed, the higher the crime rate. Surely, however, one does not control the other. Rather, they share something in common—the time of the year. In the summer, when it is warmer, there is both more crime and more consumption of ice cream. While such correlational findings may be interesting, the relationship between crime rate and ice cream consumption makes little sense given the third, very important variable of season or temperature.

Similarly, an intervention may be very successful in getting adults to lose weight, but if the program costs $4,000 per adult and in a group of 100, the average weight loss was 2.1 pounds over 6 months, does it make economic sense to consider the intervention meaningful?

With these two examples in mind, here are three conclusions about the importance of statistical significance relative to the meaningfulness of findings:

1. Statistical significance, in and of itself, is not very meaningful unless the study has a sound conceptual base that lends some meaning to the significance of the outcome.

2. Statistical significance cannot be interpreted independently of the context of the outcomes. For example, if you are the superintendent of a school system, are you willing to retain children in Grade 1 if the retention program significantly raises their standardized test scores by 0.5 point?

3. Although statistical significance is important as a concept, it is not the end-all and be-all of scientific research. If a study is designed correctly, then even null results reveal something very important. If a particular treatment does not work, for example, this is important information that others need to know.

More questions? See questions #74, #75, and #76.

What Is the Excel ToolPak, and How Can I Use It to Perform Tests of Inference?

Excel is the statistical application being used throughout *100 Questions (and Answers) About Statistics*, so we are featuring a very useful Excel tool, the Data Analysis ToolPak, as a way to quickly analyze results. Note that the ToolPak is only available for the Windows version of Excel, at least as of the 2011 version for the Mac. Also note, as we have mentioned before, that you can often use Excel functions to perform statistical analysis.

The Data Analysis ToolPak is an Excel add-in that should already be installed on the computer you are using.

Here's a generic example of how the ToolPak is used. In this case, we are examining the difference in athletes between the height of a vertical jump in the fall and again in the spring. To do this, we are using a *t*-test between dependent means.

1. Click the Data tab and then click Data Analysis. The Data Analysis dialog box will open.

2. Double-click on the procedure you want to use, which in this case is the t-Test: Paired Two Sample for Means, and you will see the t-Test: Paired Two Sample for Means dialog box.

3. Insert the range for the first and second groups.

4. Click the Labels box.

5. Specify an Output Range, or a cell where you want the output to appear. The completed dialog box for this example is shown in Figure 79.1

6. Click OK, and you will see the results of the analysis as shown in Figure 79.2, including the following (most relevant to our discussion):

   - The original data
   - The descriptive statistics for the two groups

Figure 79.1 The completed t-Test: Paired Two Sample for Means dialog box

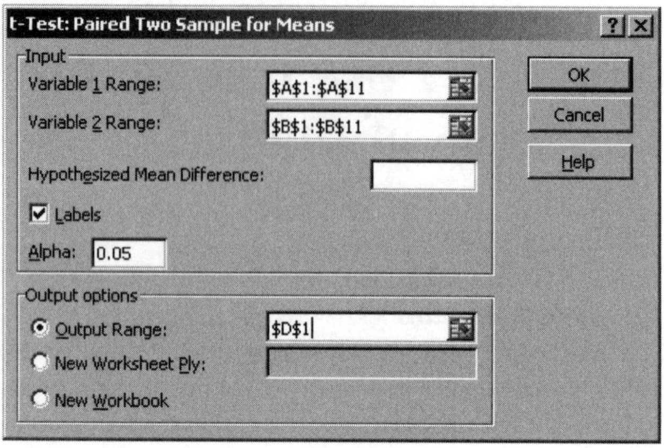

Figure 79.2 The completed ToolPak analysis

	A	B	C	D	E	F
1	Fall	Spring		t-Test: Paired Two Sample for Means		
2	18	19				
3	21	19			Fall	Spring
4	18	15		Mean	16.40	16.60
5	13	16		Variance	26.27	24.04
6	11	9		Observations	10	10
7	15	16		Pearson Correlation	0.93	
8	21	22		Hypothesized Mean Difference	0.00	
9	22	22		df	9	
10	19	20		t Stat	-0.33	
11	6	8		P(T<=t) one-tail	0.38	
12				t Critical one-tail	1.83	
13				P(T<=t) two-tail	0.75	
14				t Critical two-tail	2.26	

- The obtained t statistic, which is −0.33
- The critical value for a nondirectional two-tailed test, which is 2.26
- The probability that this value of −0.33 is due to chance, which is .75, quite removed from the conventional .01 or .05 needed for rejection of the null hypothesis. Given this information, the athletes do not differ from fall to spring in their jumping ability.

More questions? See questions #74, #77, and #80.

What Is an Excel Function, and How Can I Use It to Perform Tests of Inference?

An Excel function is a prewritten formula that performs a set of operations. For example, one of the most simple is the =AVERAGE function, where Excel returns the average or mean value for a range of cells. The following function would compute the average for the values in the cells A3, A4, and A5:

$$=AVERAGE(A3:A5)$$

Most of Excel's statistical functions (there are many other categories) can be used in conjunction with Excel ToolPak tools or on their own. Each of these ways of calculating statistical values has its advantages, so it's best you know about both.

To use an Excel function, follow these steps:

1. Select the cell in which you want the result of the function to appear.

2. Enter the function. This means typing an equal sign, the name of the function, the range of cells with the data to be analyzed, and, in some cases, other important information.

3. Click Return or Enter, and the value of the function will be returned.

For example, here's an Excel function that computes probability of the obtained t value (not the actual t value) for a test between two independent groups. The probability level is the same as the significance level. These are two sets of 10 scores. Group 1 received extra instruction, while group 2 did not.

Group 1	Group 2
5	7
6	6
3	5
4	6
6	7
5	8
4	8
6	9
5	8
7	6

You can see in Figure 80.1 that the =T.TEST function, entered in cell B13, takes the format

$$=T.TEST(A2:A11,B2:B11,2,2),$$

where

=T.TEST is the name of the function,

A2:A11 is the range of the first set of scores,

B2:B11 is the range of the second set of scores,

2 indicates it is a two-tailed or nondirectional test, and

2 means the variances in the two groups are equal.

As you can see, the resulting value of the =T.TEST function is .003, and this is the probability associated with the *t*-test value. This very low probability indicates that the difference between the means of the two groups is probably not due to chance but rather is due to the additional instruction.

More questions? See questions #74, #76, and #79.

PART 10

UNDERSTANDING DIFFERENCES BETWEEN GROUPS

How Do I Know Which Statistical Test to Use?

There are more than 100 different statistical tests that you can learn about, each one capable of analyzing data that pertain to a specific question. For example, if you wanted to find out whether there was a difference between the average scores of two unrelated groups, you would use a *t*-test for independent means. Or, if you wanted to find out whether there was a significant relationship between two variables, you would use a similar type of *t*-test but one that would use the correlation coefficient between the two variables.

It's impossible in a book like *100 Questions (and Answers)* to go into detail about every one of such a wide variety of tests, but we can provide you with a shortcut of sorts that will help you decide which test is appropriate for which type of question in a majority of the circumstances that might be presented to you.

Figure 81.1 presents a flowchart that can at least get you started on understanding which test to use under what circumstances. To use it, proceed down the chart by answering each of the questions until you get to the end of the chart. That's the statistical test you should use. A few notes about the use of this chart:

1. This flowchart does not contain all the statistical tests—only most of those you will encounter.

2. It is not a substitute for learning more about what test to use when. It's only a starting point.

3. If you come across a particular statistical test in an article or a report and wonder why it was used, you can use this chart to find the answer.

More questions? See questions #82, #84, and #86.

Figure 81.1 A simple flowchart for determining the correct statistical test

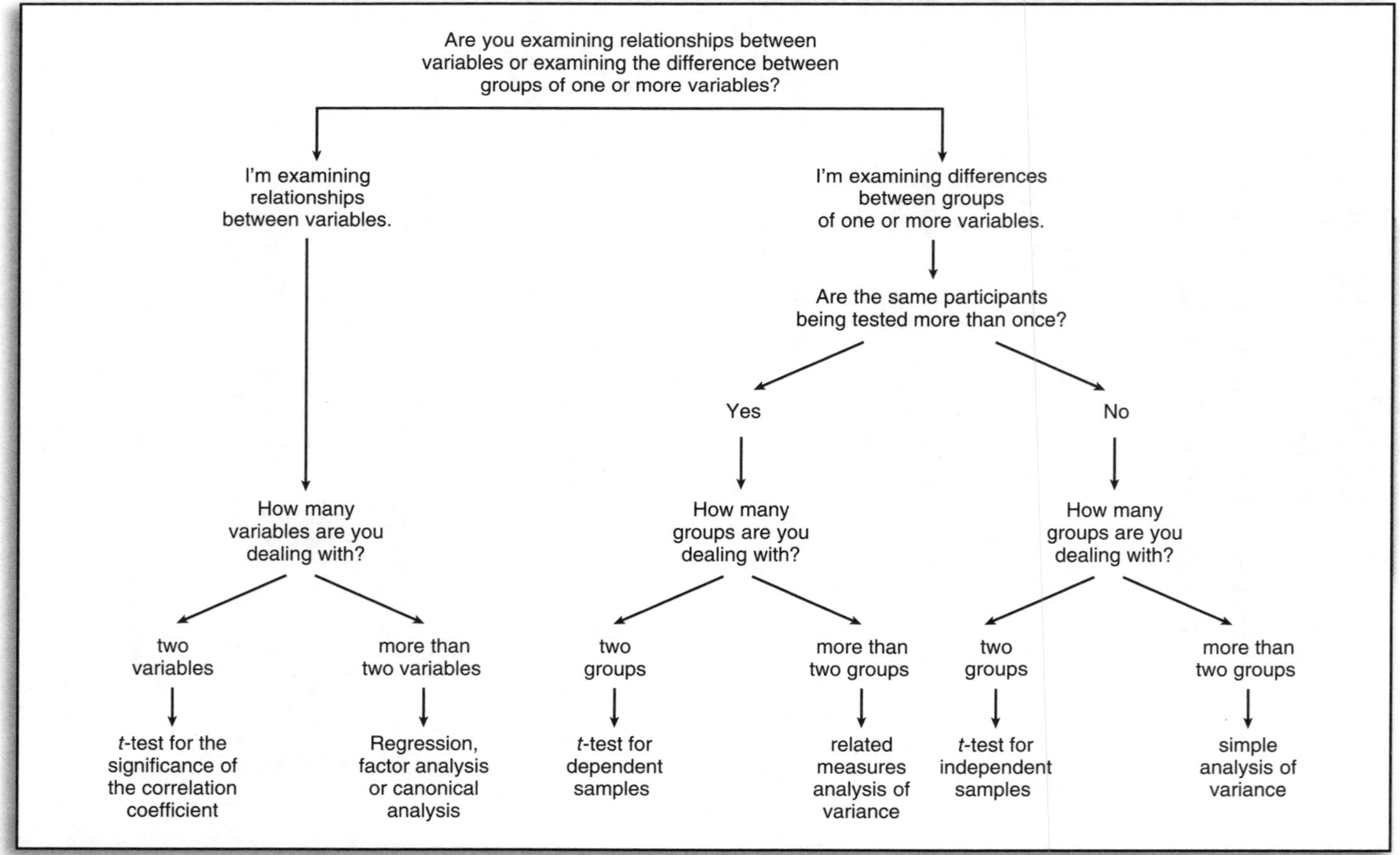

What Is a *t*-Test Between Independent Means, and What Is an Example of How It Is Used?

A *t*-test between independent means is used to test for the significance of differences between two averages from different groups. The obtained score is a function of subtracting one mean from the other and then dividing by an error term. The error term is made up of the amount of variance, or individual differences, within groups. Basically, the larger the difference between the groups, and the more similar each individual is to the others within each group, the larger the *t* value will be. A larger *t* value is more extreme and less likely to occur by chance.

Here's our example:

Eating and nutrition are very widely studied. Because of their importance, children's eating behaviors are an especially fertile field for scientists. The purpose of this study, which used the *t*-test for independent means, was to examine whether factors such as parental, peer, and media influences predict Ghanaian adolescent students' eating habits. Around 48% of the students were females and 52% males, and about 71% of them were between ages 18 and 20 years.

The scientists selected 150 students from a population of senior high school students in Ghana and asked them to complete the Eating Habits Questionnaire for Adolescents. The findings revealed a significant positive relationship between peer influence and eating habits, suggesting that the higher the peer pressure, the more unhealthy the students' eating habits.

Many other analyses were done, but for our illustrative purposes here, let's look at whether female adolescents demonstrated more unhealthy habits than did male adolescents. An independent-samples *t*-test for the two independent groups was used. The results show that there was a significant difference between the female and male adolescents. Thus, the hypothesis that female adolescents have unhealthy eating habits compared with their male counterparts is supported.

Here's the original reference . . .

Amos, P. M., Intiful, F. D., & Boateng, L. (2012). Factors that were found to influence Ghanaian adolescents' eating habits. *SAGE Open* (October–December), 1–6. doi: 10.1177/2158244012468140

More questions? See questions #81, #83, and #84.

How Can I Use Excel to Test for the Difference Between Independent Means?

There are many ways to use Excel to test for the differences between the means of two unrelated groups. Two options are the =T.DIST and =T.TEST functions. However, the most direct and easiest way is through use of the Data Analysis ToolPak.

We are using the following data to test the difference between the means of two independent groups. Group 1 was enrolled in a 2-year course with no training, and group 2 was enrolled in a 2-year course with supplemental training. The outcome variable of interest is customer satisfaction ratings (ranging from a score from 1 to 10, with 10 being best) once the individuals are on the job. The hypothesis is simply that the training (group 2) will result in significantly increased scores over no training (group 1). Here are the data.

Without Training	With Training
6	6
4	8
7	7
8	5
6	9
9	9
8	8
9	7
8	8
7	9

Follow these steps:

1. In Excel, click the Data tab and then the Data Analysis tab.

2. In the Data Analysis dialog box, click the t-Test: Two-Sample Assuming Equal Variances.

3. Enter the cell ranges and other information as you see in Figure 83.1.

4. Click OK, and you will see the output shown in Figure 83.2.

Figure 83.1 The t-Test: Two-Sample Assuming Equal Variances dialog box

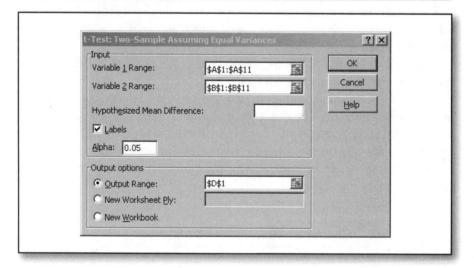

Figure 83.2 Using the Data Analysis ToolPak to perform a t-test between independent means

	A	B	C	D	E	F
1	Without Training	With Training		t-Test: Two-Sample Assuming Equal Variances		
2	6	6				
3	8	4			Without Training	With Training
4	7	7		Mean	7.6	7.2
5	5	8		Variance	1.8	2.4
6	9	6		Observations	10	10
7	9	9		Pooled Variance	2.1	
8	8	8		Hypothesized Mean Difference	0	
9	7	9		df	18	
10	8	8		t Stat	0.6	
11	9	7		P(T<=t) one-tail	0.27	
12				t Critical one-tail	1.73	
13				P(T<=t) two-tail	0.55	
14				t Critical two-tail	2.10	

As you can see, the computed t value is –0.6, and the associated probability or significance level is .27 for a one-tailed (directional) test, far beyond the .05 level. This result indicates that it is highly unlikely that the supplemental training makes a difference in customer satisfaction.

More questions? See questions #81, #82, and #84.

What Is a *t*-Test Between Dependent Means, and What Is an Example of How It Is Used?

A *t*-test between dependent means is used to test for the significance of differences between two averages from the same group. The obtained score is a function of subtracting one mean from the other and then dividing by an error term. This error term is made up of the amount of variance, or individual differences, within the same group across two different situations. The situations, for example, could be an assessment in September and one in May, two different treatments, or even two different forms of the same test. The key is that the same individuals are being tested. Basically, the larger the difference between the two assessments, and the more similar each individual's performance across testings, the larger will be the *t* value. A larger *t* value is more extreme and less likely to occur by chance.

Here's our example:

Effective reading is one of the skills that distinguishes successful students from unsuccessful students. Researchers in this area often focus on how learning to read happens and what the best interventions might be to ensure that students learn to read effectively.

In this study, two approaches to systematic word review were used in an 18-week program of extended vocabulary instruction with kindergarten students from urban schools. One approach was to teach words through extended instruction without systematic review, and the other was to teach words through extended instruction with systematic review. The primary research question was whether there was a difference in word learning between the two treatments. The study involved 80 kindergarten children who received instruction under both conditions. Hence, the scores were dependent upon each other (for each child), and the appropriate test of the research question was a *t*-test for dependent means.

As the results turned out, systematic review resulted in an almost two-fold increase in target word learning. Embedded review was effective and time efficient, whereas semantically related review was time intensive but resulted in higher levels of word learning. There was a significant gain in scores on the Peabody Picture Vocabulary Test, one of the primary measures of the effectiveness of the treatment.

Here's the complete reference . . .

Zipoli, R. P., Jr., Coyne, M. D., & McCoach, B. D. (2011). Enhancing vocabulary intervention for kindergarten students: Strategic integration of semantically related and embedded word review. *Remedial and Special Education, 32*(2), 131–143.

More questions? See questions #81, #82, and #85.

How Can I Use Excel to Test for the Difference Between Dependent Means?

We are using the following example to examine the difference between the means of the same group of adults, tested at two different times. The research hypothesis is that weight lifting makes a difference to bone density. The t-test is a one-tailed test at the .05 level of significance. These 15 female adults all participated in a weight-training program, with the initial assessment of bone density taking place in the fall and a second assessment taking place in the spring. It is expected that the second testing will show an increase in bone density. The dependent or outcome variable is a rating of bone density from 1 to 5, with 5 being the most dense. All of the participants received the same program of weight training and two bone density assessments.

Here are the data:

Fall Assessment	Spring Assessment
3	5
2	3
2	4
3	4
4	3
3	2
3	3
2	2
1	2
2	3

Follow these steps:

1. In Excel, click the Data tab and then the Data Analysis tab.
2. In the Data Analysis dialog box, click the t-Test: Paired Two Sample for Means.
3. Enter the cell ranges and other information as you see in Figure 85.1.
4. Click OK, and you will see the output shown in Figure 85.2.

Figure 85.1 The t-Test: Paired Two Sample for Means dialog box

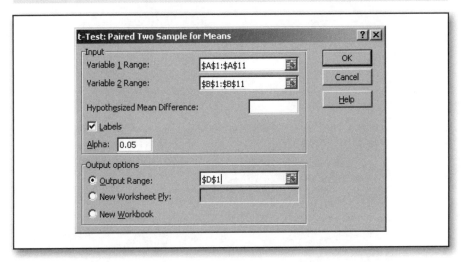

Figure 85.2 Using the Data ToolPak for a *t*-test between dependent means

	A	B	C	D	E	F
1	Fall Assessment	Spring Assessment		t-Test: Paired Two Sample for Means		
2	3	5				
3	2	3			*Fall Assessment*	*Spring Assessment*
4	2	4		Mean	2.5	3.1
5	3	4		Variance	0.72	0.99
6	4	3		Observations	10	10
7	3	2		Pearson Correlation	0.33	
8	3	3		Hypothesized Mean Difference	0	
9	2	2		df	9	
10	1	2		t Stat	-1.77	
11	2	3		P(T<=t) one-tail	0.06	
12				t Critical one-tail	1.83	
13				P(T<=t) two-tail	0.11	
14				t Critical two-tail	2.26	
15						

As you can see, the computed t value is -1.77, and the associated probability or significance level for a one-tailed direction test is .06, not quite reaching the criterion of statistical significance. Thus, the researchers concluded that the weight training did not have an effect on bone density.

More questions? See questions #80, #81, and #84.

What Is Simple Analysis of Variance, and What Is an Example of How It Is Used?

A simple analysis of variance (or ANOVA) is used to test for the significance of differences between two or more means from the same or different groups. These differences are called main effects. The analysis is termed *simple* because it tests only one dimension or factor. The obtained F value is a ratio. The numerator, or top number in the ratio, reflects the magnitude of differences between groups. The denominator, or bottom number in the ratio, reflects the magnitude of variability within each group. As the difference between means gets larger or the variability within groups gets smaller, the F ratio increases and becomes less likely to occur by chance.

Here's our example.

Most educators and parents recognize the importance of learning to read at an early age. One of the most common types of interaction between parents and their young children is picture book reading, with alphabet books being one of the most popular types of book (think, "A is for apple," "B is for bus," . . .).

This research reports the results of two studies that examined alphabet letter learning by 36-month-old children in interactions with an adult. Each child was read a standard type of children's book (nothing special about the format or presentation), a 2-D formatted book, or a book that lent itself to physical manipulation (for example, by having specially formatted pages with such features as levers).

Using a form of simple analysis of variance (testing the differences among the mean performance for the three book conditions), the researchers found that the children learned more letters with the plain books than with books with manipulative features. The researchers also found that a book feature that was specifically designed to attract children's attention to the letters did not affect performance.

Here's the complete reference . . .

Chiong, C., & DeLoache, J. S. (2013). Learning the ABCs: What kinds of picture books facilitate young children's learning? *Journal of Early Childhood Literacy,* *13*(2), 225–241.

More questions? See questions #81, #87, and #88.

How Can I Use Excel to Calculate Simple Analysis of Variance?

A nalysis of variance produces a robust test for the difference between means of two or more groups.

We are using the following data to test the differences among the means of three independent groups of police officers undergoing training on firearm accuracy. Group 1 was given no extra training; Group 2, 10 hours extra training per session; and group 3, 20 hours extra training per session. The dependent or outcome variable is accuracy, with a perfect score being 100. The hypothesis is that there is a significant difference among all three groups. Since ANOVAs or F-tests are robust and do not look at differences between pairs of means, this is not an analysis for looking at differences between specific pairs of groups. For that, post hoc analysis is appropriate.

This illustration uses the Anova option from the Data Analysis ToolPak, but you could also use functions such as =F.TEST and =F.DIST to get some of the same information.

Here are the data:

No Training	10 Hours of Training	20 Hours of Training
56	56	87
48	79	89
63	71	99
71	86	92
86	69	78
72	88	61
48	75	87
78	57	80
74	89	79
59	77	76

Follow these steps:

1. In Excel, click the Data tab and then the Data Analysis tab.

2. In the Data Analysis dialog box, click the Anova: Single Factor option.

3. Enter the cell ranges and other information as you see in Figure 87.1.

4. Click OK, and you will see the output in Figure 87.2.

Figure 87.1 The Anova: Single Factor dialog box

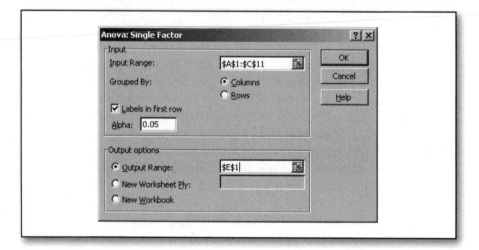

Figure 87.2 Using the Data ToolPak for the difference between three means using a simple ANOVA

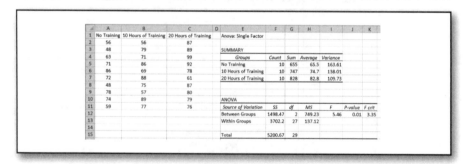

As you can see, the computed *F* value is 5.46, and the associated probability or significance level is .01—well within the region of not accepting the null hypothesis that the three means are equal. In fact, the differences among the three means of the three groups are significant.

More questions? See questions #86, #88, and #89.

What Is Factorial Analysis of Variance, and What Is an Example of How It Is Used?

While simple analysis of variance examines differences between groups on a single factor, factorial analysis of variance examines more than one factor at a time. For each factor examined, a main effect for all factors and an interaction effect are tested.

For example, a simple analysis of variance would look at differences in average language skills among third, fifth, and seventh graders. Factorial analysis of variance, on the other hand, might examine average language skills among the same three groups as well as gender. Three levels of language skills and two levels of gender (male and female) would comprise a 2 × 3 research design. The number of factors and the number of levels within those factors are only limited by the scope of the research questions and the resources available to conduct larger-scale studies. Analysis of variance produces a robust test for the difference between means of two or more groups.

In the study we are using here as an illustration, researchers examined the effects of specific practice strategies on university string players' performance. Each orchestra member was assigned to one of four treatment groups, including free practice, playing slowly before gradually speeding up, repeating small sections, and playing the excerpt multiple times. This was factor 1. Factor 2 was where each participant took both a pretest and a posttest. Thus, this was a two-factor design with four levels of factor 1 and four levels of factor 2.

No differences were found among practice strategies with regard to pitch, rhythm, expression, or overall scores. However, a significant main effect was found for test conditions: Scores were higher on the pretest.

Finally, there was no interaction between testing location and strategy. Here's the complete reference . . .

Sikes, P. L. (2013). The effects of specific practice strategy use on university string players' performance. *Journal of Research in Music Education, 61*(3), 318–333.

More questions? See questions #86, #87, and #89.

How Can I Use Excel to Calculate Factorial Analysis of Variance?

We are using the following example to examine whether there are differences between the means for two factors: gender (male and female adults are represented by rows in the below data) and type of program (program 1, program 2, and program 3—represented by columns in the data). The outcome or dependent variable is a score from 1 to 10 on a language skills test, with 10 being the highest possible score. There are two research hypotheses that Excel can test. One is that there is a difference between males and females, and the other is that there is a difference in language skills as a function of the type of program in which the participants enrolled. Both differences are being tested at the .05 level of significance.

Here are the data.

	Program 1	Program 2	Program 3
Male	6	5	8
	5	6	7
	6	5	6
	5	4	6
	4	3	5
	5	4	6
	6	5	7
Female	7	8	7
	5	7	10
	6	6	8
	7	7	6
	6	7	7
	4	9	6

Follow these steps:

1. In Excel, click the Data tab and then the Data Analysis tab.

2. In the Data Analysis dialog box, click the Anova: Two Factor Without Replication option.

3. Enter the cell ranges and other information as you see in Figure 89.1.

4. Click OK, and you will see the output in Figure 89.2.

Figure 89.1 Anova: Two Factor Without Replication option dialog box

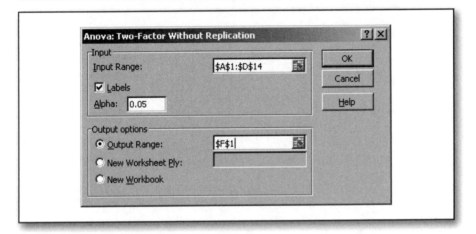

Figure 89.2 Using the Data ToolPak for a two-factor ANOVA without replication

	A	B	C	D	E	F	G	H	I	J	K
1		Program 1	Program 2	Program 3		ANOVA					
2	Male	6	5	8		Source of Variation	SS	df	MS	F	P-value
3		5	6	7		Rows	33.44	12	2.79	2.02	0.07
4		6	5	6		Columns	12.15	2	6.08	4.40	0.02
5		5	4	6		Error	33.18	24	1.38		
6		4	3	5							
7		5	4	6		Total	78.77	38			
8		6	5	7							
9	Female	7	8	7							
10		5	7	10							
11		6	6	8							
12		7	7	6							
13		6	7	7							
14		4	9	6							

As you can see, there are two computed F values. The first ($F = 2.02$) is for rows, which represent gender. The value of 2.02 is not significant beyond the .05 level, indicating that there is no difference between male and female adults in language skills. The second ($F = 4.40$) is for columns, which represent type of program. This value is significant beyond the .05 level, indicating that there is an overall difference among the three programs. Excel does not test for interaction effects.

More questions? See questions #86, #88, and #91.

How Can I Use Nonparametric Tests to Test for Significance?

Most of the questions in these last few sections of *100 Question (and Answers) About Statistics* have dealt with parametric tests or those that are based on a relatively large sample. But many situations call for nonparametric tests—tests on populations that are not adequate in size or have other characteristics that lend themselves only to nonparametric procedures.

What follows is a summary table of some of the more important nonparametric procedures, including the name of the test, when it is used, and an example.

Test Name	When the Test Is Used	Sample Research Question
McNemar test for significance of changes	To examine "before and after" changes	How effective is a phone call to undecided candidates in influencing their vote on a particular issue?
Fisher's exact test	To compute the exact probability of outcomes in a 2 × 2 table	What is the exact likelihood of getting six heads when tossing six coins?
Chi-square one-sample test	To determine whether the number of occurrences across categories is random	Did brands Fruities, Whammies, and Zippies each sell an equal number of units during the recent sale?
Kolmogorov-Smirnov test	To see whether scores from a sample came from a specified population	How representative is a set of judgments of a sample of children of the judgments of all the children at the elementary school they attend?

(Continued)

(Continued)

Test Name	When the Test Is Used	Sample Research Question
The sign test, or median test	To compare the medians from two samples	Is the median income of people who voted for candidate A greater than the median income of people who voted for candidate B?
Mann-Whitney U test	To compare two independent samples	Did the transfer of learning, measured by number correct on a test, occur faster for group A than for group B?
Wilcoxon rank test	To compare the magnitude as well as the direction of differences between two groups	Is preschool twice as effective as no preschool experience for developing children's language skills?
Kruskal-Wallis one-way analysis of variance	To compare the overall difference between two or more independent samples	How do rankings of supervisors differ among four regional offices?
Friedman two-way analysis of variance	To compare the overall difference between two or more independent samples on more than one dimension	How do rankings of supervisors differ as a function of regional office and gender?
Spearman rank correlation coefficient	To compute the correlation between ranks	What is the correlation between class rank in the senior year of high school and class rank during the freshman year of college?

More questions? See questions #81, #86, and #91.

What Is Effect Size, and
Why Is It Important?

You already know that there is an important difference between statistical significance and the meaningfulness of the results of an analysis (see question #78). But there is another very useful way to judge the value of a statistical finding, and that's through the use of effect size.

Effect size is a measure of the magnitude (not necessarily the absolute size) of a statistical finding. In other words, there may indeed be a difference between the averages of two samples, but the effect size might be very low and, therefore, the differences relatively meaningless. On the other hand, there may be a small absolute difference between the averages of two groups, but the effect size could be very large, indicating that the difference probably holds great meaning and value given the research question.

And since effect size does not take into account sample size, it provides us with yet another tool to make decisions about, for example, the importance of differences between groups.

Effect size is simple to compute. The formula is as follows:

$$ES = \frac{\overline{X}_1 - \overline{X}_2}{SD},$$

where

ES = effect size,

$\overline{X}_1$ = the mean of the first group,

$\overline{X}_2$ = the mean of the second group, and

SD = the standard deviation from either group.

For example, let's say we have the following information about a group of adolescents who took a self-esteem test and wanted to compute the effect size.

Group	Mean	s
Younger adolescents	27.5	4.65
Older adolescents	31.2	3.98

Plugging these numbers into the above formula, we have

$$ES = (31.2 - 27.5)/4.65 = 0.79$$

Interpreting the effect size is fairly straightforward. If the difference between the groups is zero, then the effect size is zero as well, and there is little difference between the distribution of scores—the scores are very similar. If the effect size is 1, then the sets of scores have about a 45% overlap, but 55% of the effect represents difference. As ES increases, the scores have less in common and are more different from one another. The larger the effect size, the more the distributions differ, and the more meaningful the difference is.

Not many researchers report effect size, but it is a very valuable tool and one effective way to further explore differences, regardless of their significance.

More questions? See questions #78, #82, and #84.

PART 11

LOOKING AT RELATIONSHIPS BETWEEN VARIABLES

How Are Relationships Tested for Significance Using the Correlation Coefficient, and What Is an Example of How the Correlation Coefficient Is Used?

Relationships between variables are tested for significance using a *t*-test, and as with any test of the significance of a statistic, the obtained value is compared to the critical value to determine whether a statistically significant relationship is present. If you refer to Figure 45.3, you can see how you can examine whether a relationship exists between two variables.

Autism, now termed part of a more broadly defined autism spectrum, is a disorder that affects an increasing number of children as well as adults. In this study, researchers examined the relationship between multisensory dysfunction—auditory, visual, touch, and oral sensory dysfunction—in autism and and severity of autism—a perfect setting for the use of correlational analysis.

The Sensory Profile was completed on 104 participants (ages 3 to 56 years) with a diagnosis of autism. Analysis of the results showed a significant correlation among the different components of the profile. An examination of the different age groups suggested that sensory disturbance correlated with severity of autism in children but not in adolescents and adults. The researchers suggested that all the main modalities and multisensory processing appeared to be affected in the participants and that sensory-processing dysfunction in autism is thus global in nature.

Here's the complete reference . . .

Kern, J. K., Trivedi, M. H., Grannemann, B. D., Garver, C. R., Johnson, D. G., Andrews, A. A. . . . Schroeder, J. L. (2007). Sensory correlations in autism. *Autism, 11*(2), 123–134.

More questions? See questions #40, #46, and #47.

How Can I Use Excel to Test for the Significance of the Correlation Coefficient?

We are using the following example to examine whether the correlation between number of hours spent using social media per week is significantly related to level of self-rated job satisfaction on a scale from 1 to 10, with 10 being highest.

The research hypothesis is that the two variables are positively related, and we will use a one-tailed test at the .01 level of significance. Each participating adult in this study recorded the number of hours per week he or she spent on such activities as Twitter, Facebook, etc., as well as their self-rating of job satisfaction.

Here are the data:

Social Media Hours	Job Success
22	5
23	6
15	5
7	3
21	9
14	5
15	6
22	7
20	8
19	9

Follow these steps:

1. In Excel, click the Data tab and then the Data Analysis tab.

2. In the Data Analysis dialog box, click the Correlation option.

3. Enter the cell ranges and other information as you see in Figure 93.1.

4. Click OK, and you will see the output in Figure 93.2.

Figure 93.1 The Correlation dialog box

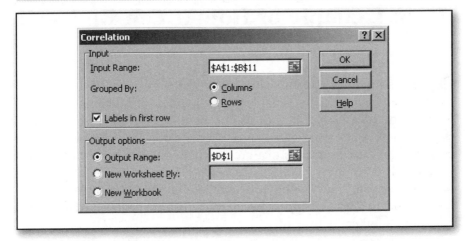

Figure 93.2 Using the Data ToolPak for correlation

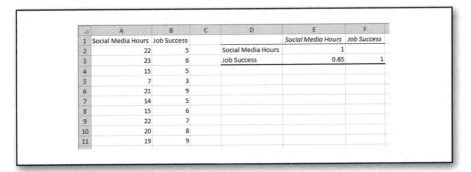

As you can see, the computed value of the correlation coefficient is .65. Excel does not yield the associated probability or significance level for a correlation coefficient, nor does the =CORREL function. To determine whether the correlation between the two variables is significant, a table of significance for this test statistics, given the number of observations, needs to be consulted. As it turns out, at the .05 level, the critical value for rejection of the null hypothesis is .54. The value of the correlation coefficient here is .65, exceeding the critical value, so our conclusion is that both variables are related.

More questions? See questions #46, #47, and #92.

What Is Simple Regression, and What Is an Example of How It Can Be Used?

Regression is a powerful technique whereby one variable can be predicted by one other variable or, in the case of multiple regression, more than one variable. In the case of multiple regression, the question asked is how well a set of variables predicts a particular outcome. The goal is to use predictor variables that are unrelated to each other but are singularly related to the variable that is being predicted. Much as in correlation, relationships are being investigated, but the goal is not just to understand the relationship between variables but to effectively predict one from another.

Obesity is a problem the world over, but despite its significance, other than overly simple explanations of eating too much and not exercising enough, scientists have relatively little idea what is at the root cause of this disease. In this study, researchers assessed various effects of nutrient intakes, health behaviors, and nutrition knowledge on the entire distribution of body mass index (BMI) across 1,491 males and 1,672 females.

The researchers found that certain factors (such as energy, oleic acid, and cholesterol) predict an increase in obesity, while other factors (such as fiber, calcium, and level of sodium) reveal a reverse effect. Their conclusions included the ideas that a high or low BMI is a risk factor to health and that BMI is a useful predictor of health outcomes.

Here's the complete reference . . .

Chen, S., & Yseng, J. (2010). Body mass index, nutrient intakes, health behaviors and nutrition knowledge: A quantile regression application in Taiwan. *Health Education Journal, 69*(4), 409–426.

More questions? See questions #46, #92, and #95.

How Can I Use Excel to Calculate a Simple Regression Equation?

In the following example, we'll examine whether the number of college football wins last season predicts the number of college football wins for the coming season. The research hypothesis is that last season's wins do predict next season's, and a one-tail or directional hypothesis is tested at the .05 level of significance.

The data, which are shown below, consist of the wins for 10 teams last season (the X variable) and those for this season (the Y variable). The correlation between the two can be used to generate a regression equation that can predict wins for the next season (which will be the Y' or Y-prime variable). We'll deal here only with testing this hypothesis concerning prediction.

Here are the data:

School	Wins Last Season	Wins This Season
1	7	8
2	6	7
3	8	9
4	11	10
5	12	9
6	8	7
7	7	12
8	3	3
9	6	5
10	5	6

Follow these steps:

1. In Excel, click the Data tab and then the Data Analysis tab.

2. In the Data Analysis dialog box, click the Regression option.

3. Enter the cell ranges and other information as you see in Figure 95.1.

4. Click OK, and you will see the output in Figure 95.2.

Figure 95.1   The Regression option dialog box

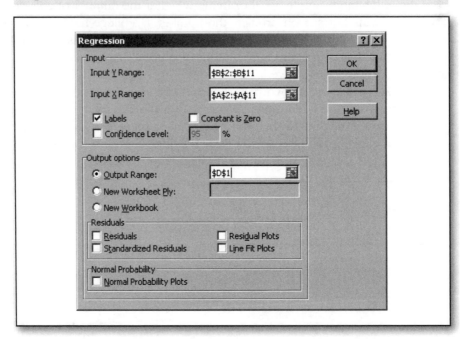

Figure 95.2   Using the Data ToolPak to calculate regression

	A	B	C	D	E	F	G	H	I
1	Wins Last Season	Wins This Season		SUMMARY OUTPUT					
2	7	8							
3	6	7		*Regression Statistics*					
4	8	9		Multiple R	0.68				
5	11	10		R Square	0.46				
6	12	9		Adjusted R Square	0.39				
7	8	7		Standard Error	2.02				
8	7	12		Observations	10				
9	3	3							
10	6	5		ANOVA					
11	5	6			df	SS	MS	F	*Significance F*
12				Regression	1	27.78	27.78	6.81	0.03
13				Residual	8	32.62	4.08		
14				Total	9	60.40			

The one computed F value, which tests the power of prediction of Y from X, is 6.81, and it is significant at the .05 level. This means that last year's wins are significant predictors of this year's (and future) wins. You can also see the multiple R value, which is .68. In this case, this value is the same as the simple paired correlation because only two variables are being correlated. In other analyses, many more variables could be correlated and used to predict a specific outcome.

More questions? See questions #40, #93, and #94.

PART 12

OTHER STATISTICAL PROCEDURES

What Is Multivariate Analysis of Variance (MANOVA), and How Is It Used?

You won't be surprised to learn that there are many different renditions of analysis of variance (ANOVA), each one designed to fit a particular "comparing the averages of more than two groups" situation. One of these, multivariate analysis of variance (MANOVA), is used when there is more than one dependent variable. So, instead of looking just at one outcome, one examines more than one outcome or dependent variable. What MANOVA does is to control for the relationship between outcome variables so that, if there is an effect, it is clear what impact the treatment has on what variables.

For example, Jonathan Plucker from Indiana University examined gender, race, and grade differences in how gifted adolescents dealt with pressures at school. The MANOVA analysis that he used was a 2 (gender with the two levels of male and female) × 4 (race, which included Caucasian, African American, Asian American, and Hispanic) × 5 (grade level, 8th–12th).

The multivariate part of the analysis was the five subscales of the Adolescent Coping Scale.

Using a multivariate technique, the effects of the independent variables (gender, race, and grade) could be estimated for each of the five scales, independent of one another.

Here's the complete reference . . .

Plucker, J. A. (1998). Gender, race, and grade differences in gifted adolescents' coping strategies. *Journal for the Education of the Gifted, 21*(4), 423–436.

More questions? See questions #86, #87, and #97.

What Is Analysis of Covariance (ANCOVA), and How Is It Used?

A nalysis of covariance (ANCOVA) is particularly interesting because it basically allows you to equalize initial differences between groups. Let's say you are sponsoring a program to increase running speed and want to compare how fast two groups of athletes can run a 100-yard dash. Because strength is often related to speed, you have to make some correction so that initial strength does not account for any differences at the end of the program. Rather, you want to see the effects of training with strength controlled. You would measure participants' strength before you started the training program and then use ANCOVA to adjust final speed based on initial strength.

Michaela Hynie, John Lyndon, and Ali Tardash from McGill University used ANCOVA in their investigation of the influence of intimacy and commitment on the acceptability of premarital sex and contraceptive use. They used ANCOVA with social acceptability as the dependent variable (they were looking for group differences) and ratings of a particular scenario as the covariate. ANCOVA would ensure that differences in social acceptability would be corrected using ratings, so this variable would be controlled.

Here's the complete reference . . .

Hynie, M., Lyndon, J., & Tardash, A. (1997). Commitment, intimacy, and women's perceptions of premarital sex and contraceptive readiness. *Psychology of Women Quarterly, 21*, 447–464.

More questions? See questions #86, #88, and #98.

What Is Repeated Measures Analysis of Variance, and How Is It Used?

Here's another kind of analysis of variance analysis. Repeated measures analysis of variance is very similar to any other analysis of variance that tests the means of two or more groups for differences. In a repeated measures ANOVA, there is one factor on which participants are tested more than once. That's why it's called *repeated*, because you repeat the process at more than one point in time on the same factor. For example, if you measured the weight of the same group of participants every week for a year and wanted to look at weekly differences, repeated measures ANOVA would be the appropriate analytic tool.

For example, B. Lundy and colleagues examined same-sex and opposite-sex interaction with best friends among juniors and seniors in high school. One of their main analyses was ANOVA with three factors: gender (male or female), friendship (same-sex or opposite-sex), and year in high school (junior or senior year). The repeated measure was year in high school, because the measurement was repeated across the same subjects.

Here's the complete reference . . .

Lundy, B., Field, T., McBride, C., Field, T., & Largie, S. (1998). Same-sex and opposite-sex best friend interactions among high school juniors and seniors. *Adolescence, 33*(130), 280–289.

More questions? See questions #86, #88, and #96.

What Is Multiple Regression, and How Is It Used?

You learned in question #94 how the value of one variable can be used to predict the value of another. Often, social and behavioral sciences researchers look at how more than one variable can predict another. This technique is called multiple regression.

For example, it's fairly well established that parents' literacy behaviors (such as having books in the home) are related to how much and how well their children read. So, it would seem quite interesting to look at such variables as parents' age, education level, literacy activities, and shared reading with children to see what they contribute to early language skills and interest in books. Paula Lyytinen, Marja-Leena Laakso, and Anna-Maija Poikkeus did exactly that, using multiple regression analysis to examine the contribution of parental background variables to children's literacy. They found that mothers' literacy activities and mothers' level of education contributed significantly to children's language skills, whereas mothers' age and shared reading did not.

Here's the complete reference . . .

Lyytinen, P., Laakso, M. L., & Poikkeus, A. M. (1998). Parental contributions to child's early language and interest in books. *European Journal of Psychology of Education, 13*(3), 297–308.

More questions? See questions #40, #94, and #95.

What Is Factor Analysis, and How Is It Used?

Factor analysis is a technique based on how items that are related to one another form clusters or factors. Each factor represents several different variables, and it turns out that factors are more efficient than individual variables at representing outcomes in certain studies. In using this technique, the goal is to represent those variables that are related to one another with a more general name. And assigning names to these groups of variables called factors is not a willy-nilly process—the names reflect the content and the ideas underlying how the variables might be related.

For example, David Wolfe and his colleagues at the University of Western Ontario attempted to understand how experiences of maltreatment occurring before children were 12 years old affected peer and dating relationships during adolescence. To do this, the researchers collected data on many variables and then looked at the relationships among all of them. Those that seemed to contain items that were related (and belonged to a group that made theoretical sense) were deemed factors. One such factor in this study was named Abuse/Blame. Another factor, named Positive Communication, was made up of 10 items, all of which were related to each other.

Here's the complete reference . . .

Wolfe, D. A., Wekerle, C., Reitzel-Jaffe, D., & Lefebvre, L. (1968). Factors associated with abusive relationships among maltreated and nonmaltreated youth. *Developmental Psychopathology, 10*(1), 61–85.

More questions? See questions #40, 49, and 92.

Index

⬤SAGE researchmethods

The essential online tool for researchers from the world's leading methods publisher

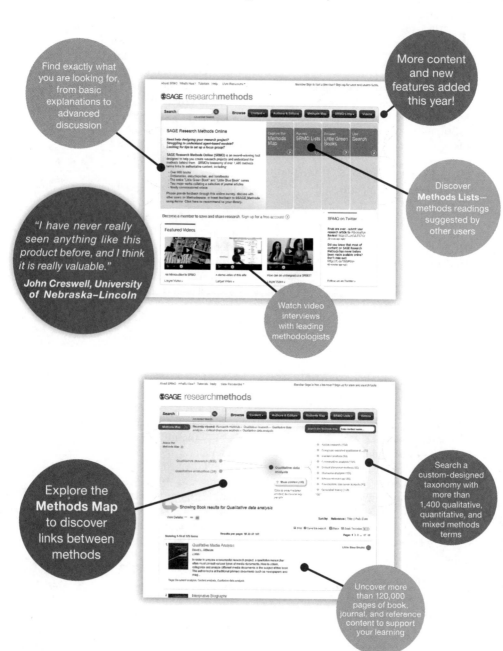

Find exactly what you are looking for, from basic explanations to advanced discussion

More content and new features added this year!

Discover **Methods Lists**— methods readings suggested by other users

"I have never really seen anything like this product before, and I think it is really valuable."

John Creswell, University of Nebraska–Lincoln

Watch video interviews with leading methodologists

Explore the **Methods Map** to discover links between methods

Search a custom-designed taxonomy with more than 1,400 qualitative, quantitative, and mixed methods terms

Uncover more than 120,000 pages of book, journal, and reference content to support your learning

Find out more at
www.sageresearchmethods.com